Classroom Activities and Experiments for Life Science

Classroom Activities and Experiments for Life Science

Lawrence E. Crum

Parker Publishing Company, Inc.
West Nyack, New York

1974

PARKER PUBLISHING COMPANY, INC.
West Nyack, New York

Library of Congress Cataloging in Publication Data

Crum, Lawrence E
Classroom activities and experiments for life science.

Bibliography: p.
1. Biology--Study and teaching (Secondary)
2. Biology--Experiments. I. Title.
QH316.5.C78 574 74-621
ISBN 0-13-136226-7

Printed in the United States of America

This book is dedicated to the study of life on our planet, in the hope that we can better understand our environment and ourselves.

How This Book Will Help You Teach Life Science

This book offers you a wealth of practical information concerning living organisms, their characteristics, and how to observe them in your secondary life science or biology classroom. You will discover unique ways to investigate the blood flowing through the wing of an insect, a *Hydra* swallowing a blood clot, or a brine shrimp emerging from its egg. You will be able to introduce your students to the unique slime mold, "water bear," gastrotrich and dragonfly larvae. You can fascinate them with *Allomyces*, the embryonic development of fish and the compound eye of *Daphnia*.

In addition to several hundred classroom activities and tested ideas, you will learn how and where to obtain specimens, in addition to the technical details necessary to set up investigations that work. You will also acquire ideas that will help you motivate students to initiate science projects.

Most of the demonstrations are inexpensive to implement and stress is placed on species having adaptability for sustained survival in the classroom. Special consideration is also given to organisms with inactive stages; these can be stored indefinitely in your classroom. Field studies are suggested as complementary or replacement activities for certain investigative areas; in general, you are provided a wide spectrum of ideas that you can adapt to your individual situation.

Whether it be through field trips or laboratory investigations, this book is designed to help you do more than simply teach life science or biology. It will help you to directly involve and interest your students, to literally bring the subject alive, to extend your classroom into the nearby community, and to motivate your students to enter and learn about living things, in *your* classroom.

In addition, some extras have been provided. You will find a list of pertinent references at the end of each chapter; these should help solve potential problems as well as provide further details for teacher and student.

A complete audio-visual listing is also provided at the end of each chapter; this includes the latest available films, filmstrips and film loops.

Following the audio-visuals is a comprehensive listing and comparison of biological supply houses. This includes a breakdown on unit prices on organisms discussed. Although prices may change, this listing provides a general idea of costs and where to go to get the most for your money.

In addition to the well-known biological supply companies, lesser-known specialist suppliers are included; this provides sources of unusual plants and animals not normally available. Such organisms often make suitable subjects for student research projects. (Endangered species are avoided.)

In summary, each chapter will provide the basics you need to introduce a host of living organisms into your classroom, in a way your students will find fascinating. The pages are filled with ideas, ideas you can build upon. It is hoped that both you and your students can discover more of the beauties and mysteries of life through *Classroom Activities and Experiments That Teach Life Science.*

Lawrence E. Crum

Acknowledgments

The ideas expressed in this book have been collected from a variety of sources. Some have been adapted and revised, and include information on integrating living organisms: where to secure them, how to maintain them, and directions you might wish to take in studying them.

Permission to include material on culture sheets, service leaflets and other materials from Carolina Biological Supply Company, Turtox, and Ward's Natural Science Establishment, Inc., is appreciated.

I'm also grateful for the assistance provided by Dr. Richard Sayre, Dr. William Kotch, Dr. Paul B. Hounshell, Grace Waters, Stewart Harris, and the Hampton Roads ETV Association.

I am especially grateful to Sheralyn Lerner for her outstanding work on the illustrations. Some have been adapted from the works of various artists, but all will help to illustrate important subject areas covered by the book.

Finally, I thank my wife, Kathy, and my children, Kevin and Kimberly, for their understanding and support.

Table of Contents

ONE

Investigating Selected Protozoa

It is easy to forget that thousands of microscopic creatures surround us at all times. It is easy to simply glance at them in a survey approach, but their small size, availability and transparency enable us to look into life itself. In addition, protozoans are available year-round and can be cultured in the classroom. Why not spend more of your time studying the fascinating protozoans (some of which are shown in Figure 1-1)?

LITTLE-KNOWN FACTS ABOUT PROTOZOA

Although protozoans are usually smaller than a millimeter across, some can be seen with the naked eye. While they appear comparatively primitive, by no means are they less than complex. In addition to possessing complicated physiological processes, from reproduction to digestion, they have adapted for survival in a diversity of habitats.

Some live exclusively in the soil; others frequent the waters of pitcher plants, live in holes of rotted trees, or exist as plankton in lakes. They may live in hot springs, the Great Salt Lake, within polluted water, or under ice-covered ponds. Due to their size and ability to form cysts, many become airborne as dust.

While most protozoans prefer neutral waters, some can thrive in extreme alkaline conditions (pH 11); others can exist in extremely acid waters (pH 2).

Their home may be their food supply as with the protozoan parasites that live within larger animals, from earthworms to man.

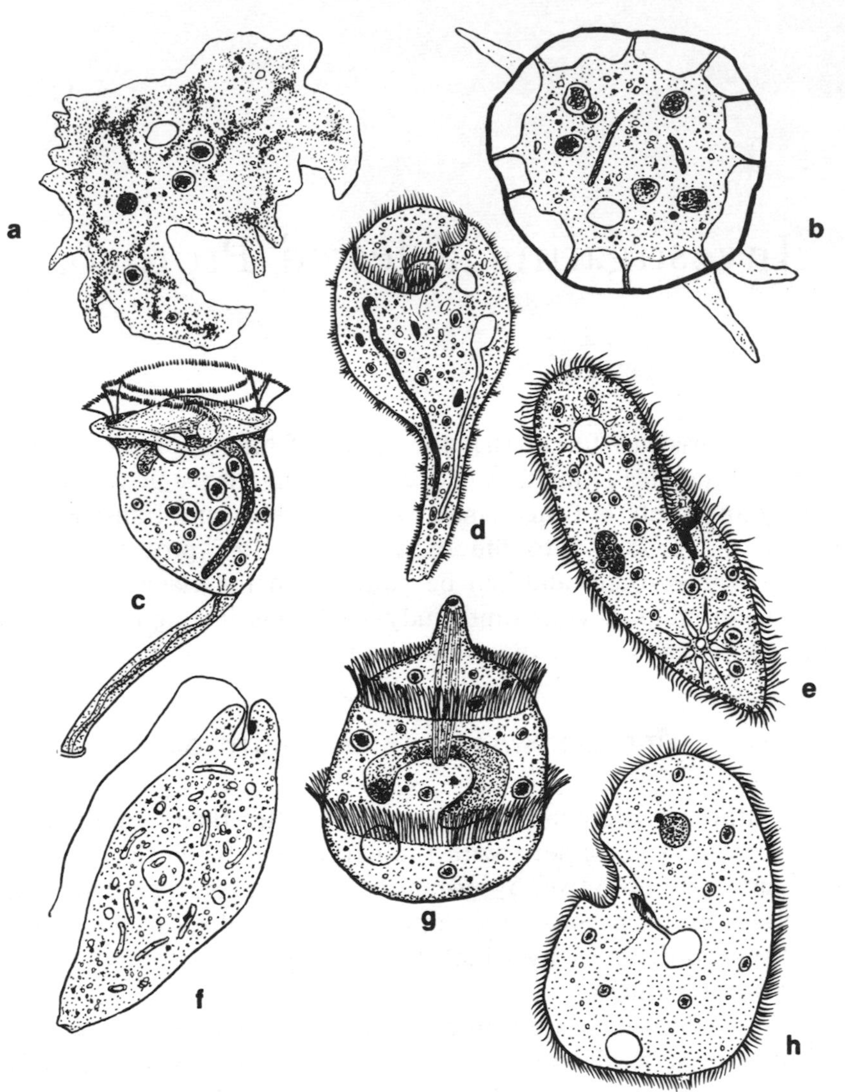

FIGURE 1-1

a. *Amoeba proteus*
b. *Arcella* (shelled amoeba)
c. *Vorticella*
d. *Stentor*
e. *Paramecium*
f. *Euglena*
g. *Didinium*
h. *Colpidium*

Original drawings by Sheralyn Lerner; after various authors; not to scale.

They may synthesize their own food or depend on other living things; they may even live both existences in their life cycles.

Their shapes are endless: cigar, trumpet, spherical, oblong, inverted bell, a changing "blob" or any number of variations. Some have spines, others have shells; some are free-swimming, others live attached; some bear cilia, others possess flagella or pseudopodia; some may utilize more than one means of locomotion in a life cycle.

Some protozoans require specific vitamins for normal growth and development; others have symbiotic relationships with algae, or associations with larger organisms.

When generalizing with protozoans we rely on broad descriptions or facts, but this does injustice to the over forty thousand individual species. To know them well, you must study them intensely.

Since protozoa exist in a diversity of habitats, you should have little trouble securing examples. However, you may wish to sustain cultures or expand the possibilities in your classroom. One way to accomplish this is to make a hay infusion.

MAKING A HAY INFUSION

A typical hay infusion is made by boiling five to ten grams of timothy hay in a quart (liter) of pond water. Allow the solution to cool overnight before removing the debris; filter, and pour into several jars or fingerbowls.

Add about two or three times as much local pond water to the containers and label the date and contents. Sample over a period of several weeks or longer.

A hay infusion does not necessarily require timothy hay. Other substitutes are dried grass, weeds, or bedding hay used in horse stalls. For other infusions, you can use the common aquarium plant *Elodea,* lettuce, pepper, bay leaves, rice, or wheat grains. (See the references at the end of the chapter.)

In general, a hay infusion can be made by mixing any number of combinations of ingredients in varied amounts of pond or distilled water. It is generally preferable to have a "weaker" solution, so an abundance of nutrients and boiling are not usually necessary. However, the ingredients are normally boiled (or warmed) and allowed to remain in the same (or other) water for 24 to 48 hours. Try your own combinations.

If you wish to stock a bacteria-free hay infusion, simply filter and autoclave (or pressure cook) the solution for 20 minutes at 15 atmospheres pressure.

If you observe a hay infusion over a period of several weeks or longer, you will notice population changes taking place, or the process of ecological succession.

ECOLOGICAL SUCCESSION IN HAY INFUSIONS

Ecological succession is a term often used to indicate an orderly sequence of changes that take place within a specific environment; the changes involve interaction between populations of organisms and environmental factors.

Just as a "typical" lake becomes dry land in a number of stages, there are certain stages that an aging hay infusion may be expected to follow. However, just as some lakes may not follow the general successional pattern, a hay infusion may not follow its predicted route. Regard succesion as a general glimpse at an often occurring process, one that is predictable but not absolute.

Studies have shown that in a freshly prepared hay infusion, very small flagellates are the first protozoans to dominate the waters. After a period of time, often within one or two weeks, small kidney-shaped ciliates dominate *(Colpoda).* These may later be replaced by other ciliates slightly larger, but differing in structure (hypotrichs such as *Euplotes*). Eventually, ostracods, or small "clam-like" shrimps, become most numerous.

It has been noted that once a particular population hits its peak it soon diminishes in numbers. There may also be vorticella, paramecium, amoeba, and rotifer organisms as well as other species appearing in the later stages of succession.

In your classroom you can easily prepare a sample hay infusion (i.e., ten grams of timothy hay or dried grass boiled in a liter [quart] of pond water for approximately ten minutes). Filter, pour into a number of containers and add a small amount of freshly obtained pond water. Have students take general counts of organisms present and diagram their appearance. (Unless you have an oil immersion microscope or lens, certain of the smaller initial species will be difficult to diagram.) After six to eight weeks discuss the results. Among the questions that could be raised are:

1. Why do changes take place?
2. Would hay infusions of different composition show the same successional patterns? (Offers a future direction for interested students.)
3. Where do the organisms come from? Why do they suddenly appear in the cultures?
4. What is the first type of organism that must be present in order for succession to take place? (Bacteria.) Why? (They provide direct or indirect food for other organisms.)
5. Does the water change in acidity during succession? (Test with PH indicator papers.)

In working with variations of hay infusions, you might consider searching for "natural infusions." Bog waters and the water in a tree hole might offer infusion variations. You might even try a rain-soaked (24-48 hours) portion of a grass field. These are but a few of the countless water sources available.

In discussing succession you may wish to relate the value of being able to predict successional changes. Are there patterns that we can expect to take place because of human overpopulation?

In your studies with ecological succession you will probably encounter a diversity of distinct protozoans. One such organism you may discover and wish to pursue further is the amoeba; a common species is *Amoeba proteus.*

Amoeba Proteus

Amoeba proteus is an excellent species for examination in the laboratory; it is available from most biological supply houses at inexpensive costs; it is a hardy species and ships well. It is also well described in many texts.

Although *Amoeba proteus* can be found on submerged leaves within larger lakes, it is advisable that pure cultures be ordered for serious study. *Amoeba proteus* is a colorless organism and is difficult to locate in the cultures you will receive. To help alleviate the problem, Carolina Biological Supply Company offers a vitachrome stain that brings a red to pink color into the amoeba. Not only does it make this organism more recognizable in cultures, but more of the cell contents are highlighted in the process. The food vacuoles appear to be the most deeply stained structures in the amoeba.

When observing this amoeba you may wish to include the following questions:

1. Is the amoeba formless or does it have anterior and posterior ends?

Through observation you should be able to record that the amoeba has a posterior or tail region (uroid), and an anterior region that advances the organism.

2. How does the amoeba "eat"?

Most likely, food organisms will be present if ordered from a supply company. Ingestion may be seen as the amoeba flows over small organisms (protozoans, bacteria, algae) and forms a watery cup; this eventually creates a bubble with the captured organism taken within the body of the amoeba. Captured organisms may be seen struggling within the bubble(s) or vacuoles. If followed, the victims will usually be killed within a few hours, then divided into smaller vacuoles, and later digested as completely as possible. (Eventually the waste portions will be excreted near the tail region.)

3. Does an amoeba "drink"?

To demonstrate this you can add a salt solution (or gelatin) to the cultures. Whitten and Flagg (1970) suggest a drop of 0.125 M NaCl in 0.01 M phosphate buffer, between 6.5 and 7.0 pH. Other researchers have used egg albumen (3 percent), rabbit gamma globulin, and a variety of chemicals to induce "drinking" (which is technically called *pinocytosis*). After adding the inducer to the culture, immediately observe it beneath the microscope. Small channels should be seen forming; these may remain for only a few moments although they sometimes persist for about half an hour. The absorbed liquids form small vacuoles within the amoeba; this appears to be an important process since about one-third of an amoeba's volume is taken in through pinocytosis every three hours. (The vitachrome stain is reported to facilitate viewing this process.)

4. How does the amoeba reproduce?

You will probably have to spend some additional time in the laboratory to witness cell division of the amoeba. A general clue to the onset of division is the spherical or near formless shape of the amoeba, with less active locomotion. It soon stretches out and pinches off in the center; the process takes about half an hour. Sometimes more than two individuals result from cell division (trinary fission?). Refer to Pendergrass and West (1969).

5. How does the amoeba move?

You should be able to describe the protoplasmic or cytoplasmic streaming in the amoeba without difficulty. Theories behind the movement are interesting. One theory suggests the change of the protoplasm from a sol to gel state, pushing the amoeba forward; another theory suggests the change of states pulls the amoeba forward. For details refer to Mackinnon and Hawes, (1961) Vickerman and Cox, (1967) and Allen (1962) in the references.

Amoeba proteus can be cultured in finger bowls, but avoid placing cultures in direct sunlight or in temperatures that vary drastically from room temperature (70-80°F.). Use wheat kernels and short cuttings of timothy hay as food; boil them briefly and add one grain of wheat and one short stalk of timothy hay to about 100 cc. of water. The biggest problem, although there are many variables, seems to be in keeping down the number of bacteria that enter the culture. If bacteria (and accompanying paramecia and other protozoans) are kept to a minimum, the amoebas have a better chance for survival. So, when you stock your food supply with amoeba, attempt to include as few bacteria as possible. If the bacteria and other protozoans increase drastically in numbers, reduce your food supply and replenish it more frequently. Change the water in the cultures at least every two weeks; remove half of the water and add distilled water in its place.

If timothy hay is scarce, you might try culturing amoebas in a wheat infusion made by adding several boiled grains of wheat to 100 ml. of distilled water. If you can locate amoebas on decayed water lily leaves in a nearby pond, you might attempt to use the leaves as a medium, but all of these methods are marginal.

Biological supply houses offer a prepared medium for culturing amoebas (see listings at the end of the chapter). Even so, other variations for culturing amoeba, with a better chance for sustained survival, include adding *Tetrahymena* as food (on a continual basis), or adding a few grains of polished rice to Chalkey's medium (NaCl, 0.1 gm; KCl, .004 gm; $CaCl_2$, .006 gm; added to 1000 ml. of distilled water). Refer to Chalkley (1930) Turtox Service Leaflet No. 4, Zipko (1971), Hawes (1961) or Wards Culture Leaflet No. 1: "Culture of Protozoa in the Classroom" for details.

In any case you will probably find that *Amoeba proteus* are not among the easiest of protozoans to culture.

When investigating *Amoeba proteus,* your students might assume that all amoebas resemble this species. It may surprise them to learn about shelled amoebas, often represented in the laboratory by the more common *Arcella.*

Arcella

In observing shelled amoebas, you have an opportunity to compare the larger blunt-tipped pseudopodia of *Amoeba proteus* to the more pointed pseudopodia of shelled amoebas. Of course the most obvious difference between the two organisms is the shell or test.

The shelled amoeba *Arcella* secretes its brown to yellow-brown shell, although other shelled amoebas often build their shells from sand grains and pond detritus.

Arcella can be found existing on submerged plants, among mosses, and within stagnant waters. Cultures are readily available from most biological supply houses (refer to listings at the end of the chapter).

When examining *Arcella* you might consider the following questions:

1. Why must cover glasses be supported when observing *Arcella?*

If cover glasses are not supported, the shells may be crushed during slide preparation; even so, there may be empty shells on the properly prepared slides.

2. How would you describe the movement of *Arcella?*

Pseudopodia extend out, attach, and pull the organism forward; there may be a rocking motion caused by the presence of the shell.

3. How does *Arcella* reproduce, with a shell that would appear to be in the way?

It is apparent that this shelled amoeba divides within its shell. The new shell-less individual emerges from the opening on the underside of the original shell; it soon creates its own shell and departs.

4. Of what value is the shell?

It offers protection to the actual amoeba, but the color of the shell may indicate the amount of iron in the water by the degree of darkness it possesses.

Arcella can be cultured in separate or a combination of infusions of wheat and hay; keep the water level very shallow.

While amoebas spend their existence crawling on the bottom of lakes and ponds, the ciliates have mobility to move through the waters. Ciliates are numerous and represent the largest group of protozoans. While over 6,000 species have been identified, few are more easily obtainable and observable than the stentors.

Stentor

Many stentors are large, sometimes ranging up to 3 mm. in length, facilitating observation and experimentation. There are a number of species, but much of the literature and texts emphasize *Stentor coeruleus.*

Stentor coeruleus is a trumpet-shaped protozoan visible to the naked eye; it is easily identified by its blue color, chain nucleus, and the typical expanded anterior end, surrounded by rows of prominent cilia.

Stentor coeruleus may be commonly found in ponds, but it is offered extensively through biological supply houses (see listings at the end of the chapter).

When examining this organism you might wish to include the following questions:

1. Does *Stentor coeruleus* have cilia (small hairs) only on its anterior feeding end?

You should be able to detect cilia on the entire body surface.

2. What is the apparent purpose of the cilia?

The primary purposes are to facilitate feeding and locomotion. You should be able to observe feeding, noting rejection of certain particles. You may also discover that the stentor may remain temporarily attached or swim freely.

3. On what does the *Stentor* feed?

Stentors feed well on *Blepharisma,* a pink ciliate that often frequents the cultures from biological supply houses. They also prefer *Chilomonas,* other protozoans, bacteria, small rotifers, and sometimes even each other.

4. How does the *Stentor* reproduce?

While it is doubtful that you will witness conjugation, you might be able to locate "strange looking" stentors that may at first resemble two organisms joined together. Actually this is probably binary fission or asexual reproduction; eventually the two sections grow apart to form two distinct individuals.

For culturing, Mackinnon and Hawes (1961) suggest boiling approximately 20 grams of wheat or barley in a quart (liter) of water. Allow the solution to stand until bacteria or ciliates appear, before introducing *Stentor.*

Other culture techniques suggest maintaining *Blepharisma* or *Chilomonas* cultures as specific food organisms to feed *Stentor.*

When searching for ciliates in pond water you will probably discover that in addition to stentors, another common ciliate is the *Vorticella.*

Vorticella

Vorticellids are usually described as stalked ciliates although they "drop" their stalk when in unfavorable conditions; the body swims away to a new location.

Vorticellids are prominent in polluted water which supplies them an ample supply of bacteria as food. Vorticellids are available from most biological supply companies.

When studying *Vorticella* you may wish to include the following questions:

1. Although you normally see *Vorticella* attached to the bottom (sessile), can it move about?

Students should be able to observe frequent contractions of the *Vorticella's* flexible stalk which gives it definite movement, although attached. When in unfavorable conditions the *Vorticella* forms cilia at the base of the body; eventually it breaks away from the stalk and swims toward better conditions, if they exist. This free-swimming stage (telotroch) is also found in reproduction. This stage may be observed in the laboratory since unfavorable conditions are often created when preparing and observing microscopic slides of protozoans.

2. How would you describe the feeding behavior of *Vorticella*?

When the cilia bring in food particles with their currents, the *Vorticella* may be seen rejecting certain particles. It has both feeding and rejection currents. The food enters the body of the vorticellid as somewhat spindle shaped; it eventually assumes the spherical shape of other protozoan food vacuoles.

3. What causes *Vorticella* to form cysts and of what value are they to the *Vorticella*?

Students might experiment with slowly evaporating water from their slides or cultures, starving the specimens, or slowly introducing higher or lower temperatures. These conditions can cause the vorticellids to encyst. To activate the cysts, introduce freshly prepared hay infusions rich in bacteria. The vorticellids should encyst within an hour. Students might experiment with the length of time (under certain conditions) that the cysts survive. Not only do the cysts allow the protozoan to survive environmental hardships; they also enable it to be carried with the winds to new locations. Students might experiment with the dispersing role of the *Vorticella's* cyst.

4. How does the *Vorticella* reproduce?

Students may discover a *Vorticella* dividing (binary fission) to form two individuals, taking half an hour. They may also observe conjugation taking place following the excysting process (where one fuses in the lower portion of another's body).

The *Vorticella* and *Stentor* provide common examples of the numerous ciliates, but perhaps the best-known ciliate is the *Paramecium.*

Paramecium

The eight recognized species of *Paramecium* are largely found in stagnant waters, from fresh to brackish.

One way to collect *Paramecium* is to fill a large jar with pond leaves and bottom detritus; allow it to stand in the laboratory for several days and sample the surface scum for *Paraceium.* While *Paramecium* is not too difficult to locate naturally, it is readily available from biological supply companies that offer four species: *P. caudatum, P. multimicronucleatum* (Figure 1-2), *P. aurelia,* and *P. bursaria.*

When observing *Paramecium* in the classroom you might wish your students to consider the following:

1. Does the *Paramecium* move toward or away from the pull of gravity?

This ciliate will collect at the top of the water's surface when in narrow glass tubing or in a test tube, demonstrating a negative response to gravity. Students can use the response to gravity as a way to isolate the *Paramecium,* or to concentrate the numbers.

2. Does the *Paramecium* feed while swimming fast?

It normally feeds only when browsing or slowly swimming.

3. How fast can a *Paracemium* swim?

Students may be able to determine speed on a slide containing metric divisions, or approximate it through experimental efforts. One estimate places the *Paramecium's* speed from 1 mm. to 3 mm. per second. For study purposes students may wish to slow down the *Paramecium.* Cotton fibers can be used to create barriers that inhibit the movement of the *Paramecium.* Methyl

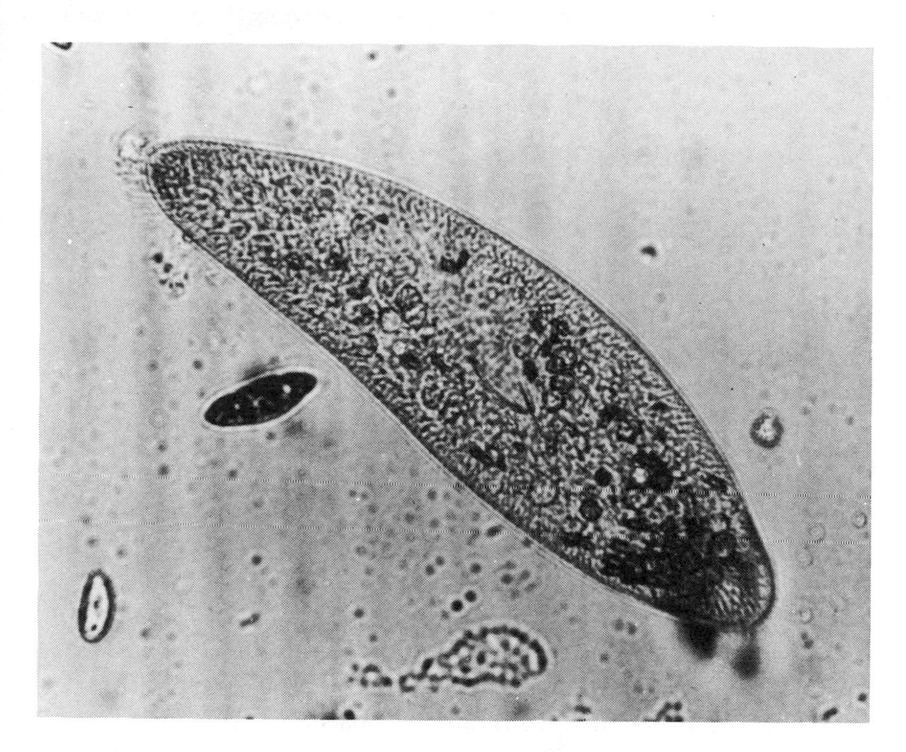

FIGURE 1-2

Paramecium multimicronucleatum **(X430) is one of four species of *Paramecium* offered commercially. Its availability, ability to ship and survive well, and comparatively larger size are a few reasons this is a popular species.**

Photo by Stewart Harris; courtesy: Hampton Roads ETV Assoc.

---

cellulose and Protoslo (see listings at end of the chapter) are two commercial preparations designed to slow down protozoans.

4. Which way does a typical *Paramecium* rotate when moving forward? When moving backward?

When moving forward the rotation is to the left; when moving backward it turns to the right. If a *Paramecium* turns to the right when moving forward you have identified a rare species, *Paramecium calkinsi.*

5. Describe the method of the *Paramecium* moving around objects.

The *Paramecium* illustrates a classic random trial-and-error movement; it continues to move forward until meeting resistance and then backs away before heading in another direction until it no longer encounters a barrier. Do humans sometimes employ the same type of response?

6. The currents that bring food into the *Paramecium,* reach ahead of the path of the organism and funnel particles toward it. What purpose could this serve?

This enables the *Paramecium* to sample conditions in the nearby waters and react in time to avoid directly meeting fatal conditions such as temperature and chemical extremes.

7. What possible means of defense does the *Paramecium* have?

Obviously, cilia provide a means to escape from some predators. Sensitivity to strong chemicals and temperature extremes offers other important survival features. Trichocysts, small hairlike spears in the edge of the membrane, appear to be of little value against predators, although they have often been suggested as defensive in function. Students can observe discharge of trichocysts by adding a drop of 0.5 percent methylene blue, ink or iodine on the slide.

8. Describe reproduction in *Paramecium.*

Students may discover fission occurring where two *Paramecia* appear joined end to end, swimming as one. Eventually the two will grow apart. *Paramecium* is well known for its primitive sexual reproduction known as conjugation. During conjugation, two strains of a species join together to exchange nuclei. While fission may occur more frequently, it appears the conjugation is necessary for the continued existence of a culture. For details of conjugation refer to Vickerman and Cox (1961) and other protozoology references. To demonstrate conjugation, order one of several conjugation kits; usually *Paramecium multimicronucleatum* and *Paramecium bursaria* are offered.

Paramecium can usually be cultured in dilute hay infusions: 1 gm. of hay per 100 ml. of distilled water. Protozoan pellets offered commercially only have to be dissolved in water to provide a suitable medium with organisms, possibly including *Paramecium.* In addition, a medium for culture of *Paramecium* is offered commercially (refer to listings at end of the chapter.)

Paramecium can be cultured, nevertheless, in wheat or rice infusions mixed with timothy hay. Add several boiled wheat or rice grains, and three times as many short (cooked) hay cuttings to a 4½-inch fingerbowl filled with distilled water. After introducing *Paramecium,* cover the container and store at room temperature (75°-85° Farenheit preferably). Avoid placing the culture in direct sunlight and add additional food when the culture decreases in numbers (in about two weeks).

Another method of culturing *Paramecium* involves slowly drying lettuce in the oven, powdering the leaves, and boiling 1.5 grams in a quart (liter) of distilled water. Filter and dilute before introducing *Paramecium.* Refer to Turtox Service Leaflet No. 4, Ward's Culture Leaflet No. 1, Zipko (1971), and Hoyt and Jablecki (1970) for details in culturing *Paramecium.*

While *Paramecium* is perhaps the best-known ciliate, the *Euglena* is perhaps the best-known flagellate.

Euglena

Often cited as half plant and half animal, the *Euglena* intrigues both teacher and student. It is a creature that can produce its own food when in sunlight, yet absorb food when in darkness.

While *Euglena viridis* is the species often cited as the typical *Euglena, Euglena spirogyra* and *Euglena gracilis* are recommended for classroom study because of larger size and ease of culturing.

You might wish to include the following questions on *Euglena:*

1. How would you describe the movement of *Euglena*?

The "tail" or flagellum of this organism moves about 12 times a second as the body gyrates and rotates. The flagellum was once thought to be of little value in movement, but now it is felt to be of significant influence in moving the *Euglena.* (Actually the *Euglena* has two flagella, one being very short.) In addition to its swimming motion, the "Euglenoid movement" may also be seen; this is where a wave passes up and down the *Euglena,* distorting the shape but giving motion in the process.

2. At the base of the flagellum is a dark pigment, sensitive to light, called a stigma or "eyespot." Of what probable value is the eyespot?

The eyespot is thought to guide the *Euglena* toward light where it can manufacture its own food with its chlorophyll (and the presence of vitamin B_1 and B_{12} which it needs to multiply).

3. How does the *Euglena* reproduce?

Students may witness binary fission, in which the *Euglena* becomes rounded and eventually splits apart, beginning at the tail region, and becomes two individuals.

Cysts may be induced by experimenting with strong lights and varying chemicals and temperatures. Cysts are thought to play important roles in the life cycles of some *Euglenas.*

Euglena has been cultured successfully in tubes of water containing numerous boiled wheat grains. Try a variation of this method by filling a large battery jar with distilled water. Add an abundance of boiled rice, wheat, or timothy hay. Place near a window and add *Euglena* gathered from stagnant pools or purchased cultures. Add more food in approximately one month.

Euglena requires little attention; it can survive wide pH differences and temperatures up to 90°F. In addition, a prepared medium can be obtained commercially. (Refer to listings at the end of the chapter).

While individual species are interesting to study, the typical fresh-water habitat is filled with a diversity of protozoans that are in close proximity to one another. While bacteria and much smaller plant-like organisms often supply food for this mixture of protozoans, certain larger species feed on larger, better-known species. In these more obvious cases, a "predator-prey" relationship is sometimes assumed.

OBSERVING PREDATOR-PREY RELATIONSHIPS IN THE MICROSCOPIC WORLD

While a predator in the microscopic world is normally a prey as well (in relationship to larger organisms), we can find certain organisms feeding on certain others. For example, *Didinium* is an efficient predator of *Paramecium.* It has adapted to ingest the *Paramecium.*

However, *Didinium* is a rapid-swimming ciliate and relatively difficult to follow under the microscope; in addition, more than one *Didinium* may attack a much larger *Paramecium.* Therefore, observing *Didinium* in action may be difficult to follow, and confusing when you do.

A more reliable feeding sequence can be seen in *Pelomyxa carolinensis–Paramecium* cultures. *Pelomyxa carolinensis* (Figure 1-3) is a giant amoeba that is sometimes called *Chaos Chaos*; because of its large size (it can be seen with the naked eye), relatively slow-flowing motion restricted to the bottom, and transparent body, this giant amoeba provides an excellent predator

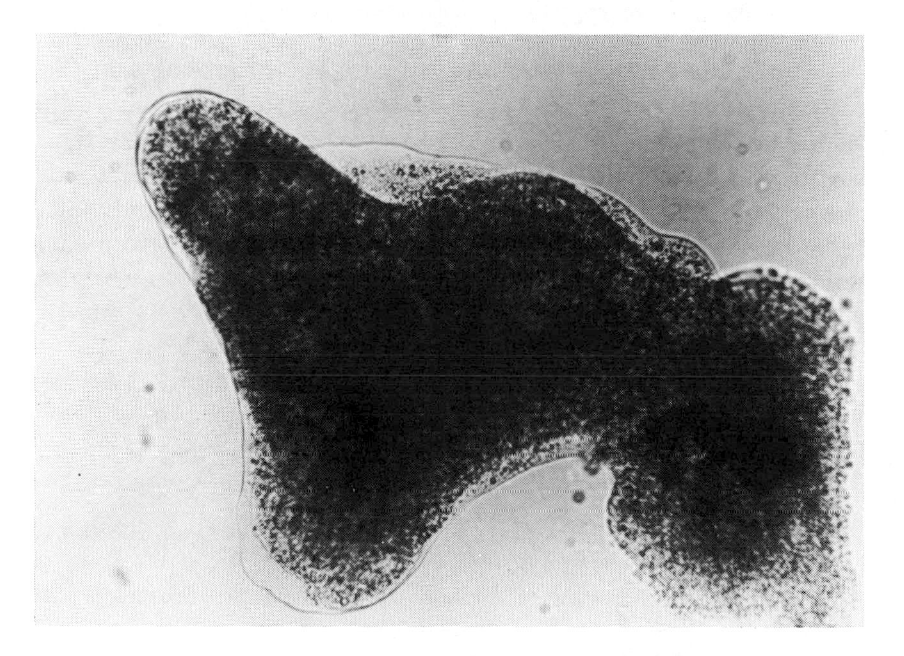

FIGURE 1-3

The giant amoeba, *Pelomyxa carolinensis* **(X100), provides an example of an active protozoan that has been termed a predator.**

Photo by Stewart Harris; courtesy: Hampton Roads ETV Assoc.

to follow under the microscope. In addition, each organism may capture a number of *Paramecium.*

Pelomyxa can be seen feeding on the green-colored *Paramecium busaria* or the pink *Blepharisma,* but perhaps the most impressive prey is the blue *Stentor coeruleus.* (Refer to listings at the end of the chapter).

Order an extra culture of prey, and a small drop of each species on one slide is all that is necessary. Feeding begins almost immediately.

Questions can include:

1. How many "prey" are captured by a single amoeba?
2. How is the "prey" captured?
3. How long can the "prey" live within the amoeba?
4. Do "prey" ever escape the amoeba?
5. How long does the ingestion process take?

In concluding this unit on protozoans, it must be said that there are countless species that have been omitted; if you have the time and facilities perhaps you can expand far beyond these suggestions. You might experiment with magnets or electrical currents or any number of chemicals in determining individual responses (For example, a six-volt battery with wires from each pole will evoke a response from *Paramecium,* if the two wires are allowed to touch the drop of water on the slide. Refer to Youngpeter, 1971.)

Use your imagination; the possibilities are infinite.

REFERENCES

Allen, R.D. "Amoeboid Movement," *Scientific American,* vol. 206, no. 2 (February 1962), pp. 112-122.

Barnes, Robert D. *Invertebrate Zoology.* Philadelphia: W.B. Saunders Co., 1968.

Burkholder, P.R. "Cooperation and Conflict Among Primitive Organisms," *American Scientist,* vol. 40, no. 4 (October 1952), pp. 601-631.

Chalkley, H.W. "Stock Cultures of Amoeba," *Science,* vol. 71, no. 1843 (April 25, 1930), p. 442.

Hairston, N.G. *et al.* "The Relationship Between Species Diversity and Stability: An Experimental Approach with Protozoa and Bacteria," *Ecology,* vol. 49 (1968), pp. 1091-1101.

Hoyt, Loris A. and Mrs. Charles Jablecki. "Paramecium–Cheap and Easy," *American Biology Teacher,* vol. 42, no. 9 (December 1970), p. 555.

Lotz, Janet and Thomas R. Mertens. "Ecologic Succession in Hay Infusions: Search for a One-Lab Presentation," *American Biology Teacher,* vol. 33, no. 2 (February 1971), pp. 94-97.

Mackinnon, Doris L. and R.S.J. Hawes. *An Introduction to the Study of Protozoa.* Oxford: Clarendon Press, 1961.

Manwell, R.D. *An Introduction to Protozoology.* New York: St. Martin's Press, 1961.

Pendergrass, William R. and W.R. West. "Fission in Amoeba," *Carolina Tips,* vol. 32, no. 8 (August 1969).

Pennak, R.W. *Fresh-Water Invertebrates of the United States.* New York: The Ronald Press Company, 1953.

Turtox Service Leaflet No. 4, "The Care of Protozoan Cultures in the Laboratory." Chicago: CCM: General Biological, Inc.

Vickerman, K. and F.E. Cox. *The Protozoa.* Boston: Houghton Mifflin Company, 1967.

Ward's Culture Leaflet No. 1, "Culture of Protozoa in the Classroom." Ward's Natural Science Establishment, Inc., Rochester, N.Y.

Whitten, R.H. and R.O. Flagg. "Vitachrome Cultures," *Carolina Tips,* vol. 33, no. 11 (September 1970).

Youngpeter, John M. "Teacher Workshop Notes," *Ward's Bulletin,* vol. 11, no. 77 (October 1971).

Zipko, Stephen J. "Protozoan Cultures in the Laboratory," *Ward's Bulletin,* vol. 11, no. 78, Rochester, N.Y. (November 1971).

*AUDIO-VISUALS**

Films:

Ameboid Organisms (AIBS) (McGraw-Hill)
Amoeba (UW and EBEC)
Life in a Drop of Water (Coronet)
Life in a Pond (Coronet)
Living Things in a Drop of Water (EBEC)
Microbiology, Part II: Complex Microorganisms, No. 3 (McGraw-Hill)
Microscope and Its Use (McGraw-Hill)
Paramecium, Euglena, and Amoeba (Wards)
The Protist Kingdom (BFA)
The Pond (IFB)
The Single-Celled Animals–Protozoa (EBEC)

Filmstrips:

Introduction to Protozoa (SVE)
One-Celled Animals (EBEC)
World of One-Celled Animals (PSP)
Zoology Series (set 1) (McGraw-Hill)

*Audio-visuals generally intended for upper elementary through high school audience; refer to glossary (page 211) for key to abbreviations and addresses of companies.

Film Loops:

Actinosphaerium (Thorne)
Amoeba (CCM and Thorne)
Arcella (Thorne)
Blepharisma (CCM)
Didinium (Thorne)
Euglena (CCM and Thorne)
Euplotes (Thorne)
Locomotion in an Amoeba (BSCS)
Microscopic Animals–Protozoa (Doubleday)
Paramecium (I-IV) (Thorne)
Protists (10 loops) (Thorne)
Raising Microscopic Animals (Doubleday)
Spirostomum (Thorne)
Stentor (CCM and Thorne)
Stylonychia (CCM)
Vorticella (CCM and Thorne)

LIVING PROTOZOA: FOR CLASS OF 12 (OR MINIMUM PRICE)*

	BICO	*Carolina*	*Conn.*	*CCM Turtox*	*So. Bio.*	*Stansi*	*Schettle (Moguled)*	*Stein. (Nasco)*	*Wards*
Amoeba proteus	2.00	2.25	2.00	2.30	2.60	2.25	3.50	2.75	3.00
Arcella	1.50	1.75	1.75	1.80	2.15	1.75	3.50	2.75	3.00
Stentor	1.50	1.75	1.50	1.85	2.15	1.75	3.50	2.75	3.00
Vorticella	1.75	2.00	1.75	2.10	2.45	2.00	3.50	2.75	3.00
Paramecium	1.50	1.50	1.75	1.80	1.95	1.50	3.50	2.75	3.00
Paramecium: conjugating	2.75	3.50	3.50	3.60	4.40	3.50	6.00	2.75	5.00
Euglena	1.50	1.50	1.50	1.60	1.95	1.50	3.00	2.75	3.00
Pelomyxa-Paramecium	2.50	3.00	2.50	3.10	3.20	3.00	3.50	2.75	3.00
Pelomyxa-Stentor	–	–	–	–	4.90	3.00	–	–	–
Pelomyxa-Blepharisma	–	–	–	–	3.50	3.00	–	–	–
Didinium-Paramecium	2.00	3.00	2.25	4.35	–	3.00	3.50	2.75	3.00

The higher costs generally indicate a class of 25. Most of these sources supply four species of *Paramecium;* the indicated price usually indicates the larger *P. multimicronucleatum* species. Two conjugating *Paramecium* species are offered: *P. multimicronucleatum* and *P. bursaria.*

The prices indicated were in effect at the time this book was written. It is suggested that when you write for cultures, you request the organization's current price list.

---

*Refer to glossary (page 210) for key to abbreviations and addresses.

While Edmund Scientific Company is not a biological supply company, they do offer (for class of 25 minimum):

Amoeba proteus	$4.50
Euglena	3.00
Paramecium	3.25
Paramecium, conjugating	5.50
Quieting solution	1.50 (1 oz.)
Dried protozoan papers	$1.50

Address: Edmund Scientific Company
615 Edscorp Building
Barrington, New Jersey 08007

Protozoa Media and Accessories

Methyl Cellulose	Offered by most biological supply houses; $1.00 per 4 oz.
Protozlo	Offered by Carolina Biological for $1.00. (Similar type of material offered by a few other companies.)
Copper Sulfate and Copper Acetate	Offered by Southern Biological and BICO for slowing down ciliates. Generally priced higher than methyl cellulose or protoslo.
Preheated Rice or Wheat Grains	Offered by most supply houses for approximately $1.00 per 4 oz.
Preheated Timothy Hay	Offered by a variety of supply houses. Prices range from $.75 to $1.80 per 4 oz.
Paramecium medium	Offered by several companies. Price range between $2.25 to $4.00, for 1000 ml.
Amoeba medium or Euglena medium	Offered by several companies. Price range between $2.25 to $4.00, for 1000 ml.
Infusoria powder	Offered by CCM and BICO. Give rise to *Arcella, Euglena, Didinium,* and others when dissolved in water. Cost range: $1.00 to $1.75.

TWO

Invertebrates Beneath the Microscope

There are few more impressive ways to discover what life is all about than through viewing aquatic invertebrates beneath a microscope. Not only do they provide representatives of larger, more difficult to handle land invertebrates, but they offer a view into life itself. Where else can you watch an organism's beating heart, flowing blood, wave-like intestinal movement, or the manipulation of muscles? Yet these things do not require surgery. All you need is a microscope and you can discover some of the upcoming organisms, and much more. (See Figure 2-1.)

HOW TO LOCATE AND INVESTIGATE THE UNIQUE TARDIGRADE (WATER BEAR)

A tardigrade is a microscopic invertebrate that is different from all others. It has a body composed of a distinct head and four body segments with four pair of legs; two pair of claws are found at the end of each leg (Figure 2-2). Its clawing or pawing motions make it resemble a minature bear, from which fact it receives its common name.

You will find a water bear's physical appearance is only part of its uniqueness (Figure 2-3). Water bears live almost exclusively in the film of water that covers mosses, liverworts and lichens. They have adjusted to living through periodic dry spells.

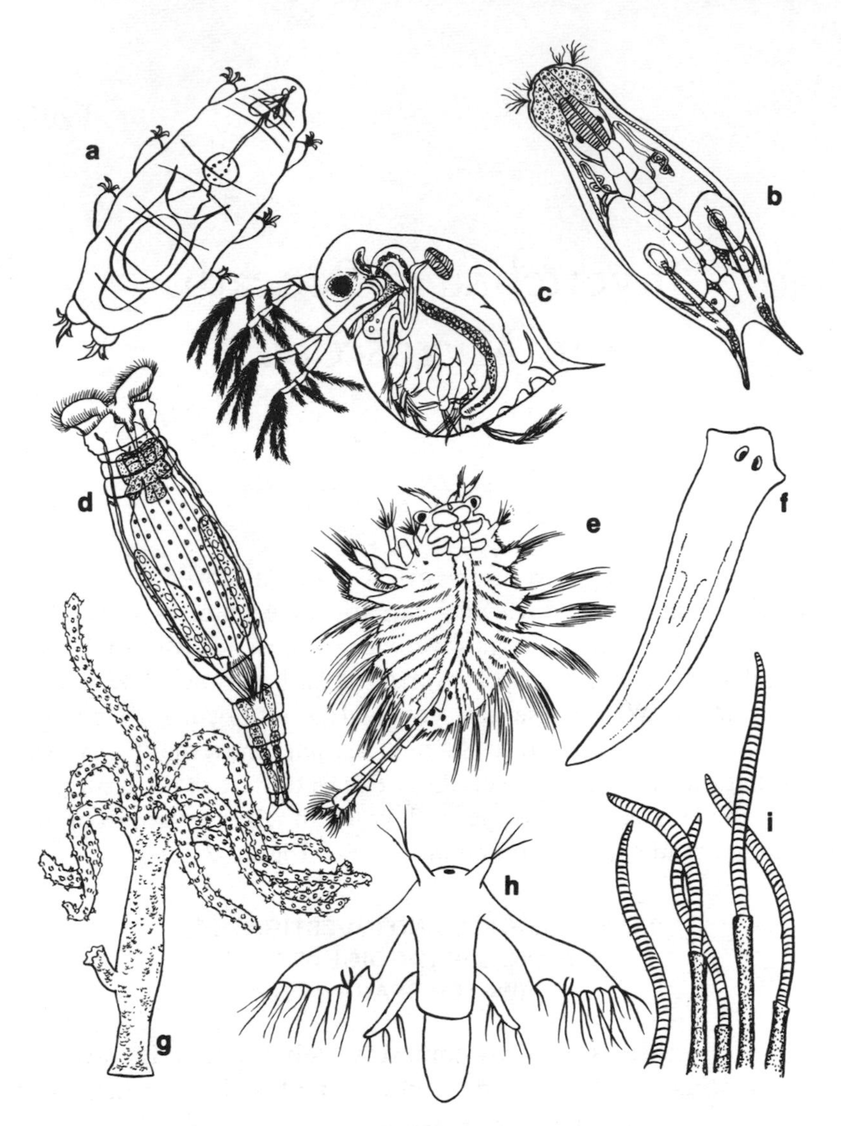

FIGURE 2-1

a. water bear
b. gastrotrich
c. water flea *(Daphnia)*
d. rotifer
e. brine shrimp, adult
f. planarian
g. *Hydra,* emphasizing stinging cells and bud
h. larva (nauplius) of brine shrimp
i. *Tubifex* worms

Original drawings by Sheralyn Lerner; after Pennak and various authors; not to scale.

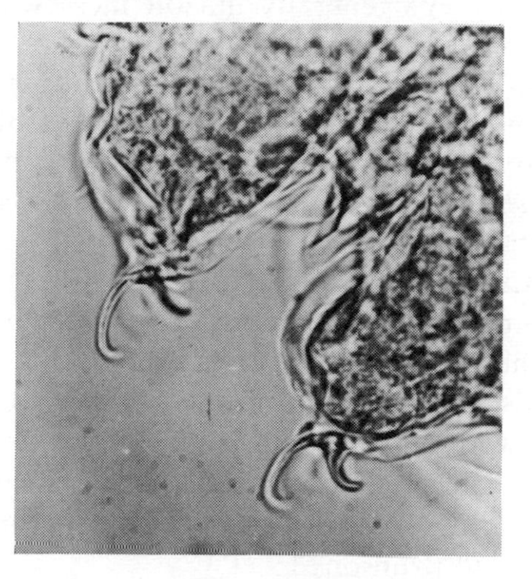

FIGURE 2-2

The claws of a water bear can be retracted. They are one of the features that identify the water bear.

Photo courtesy: Richard Sayre

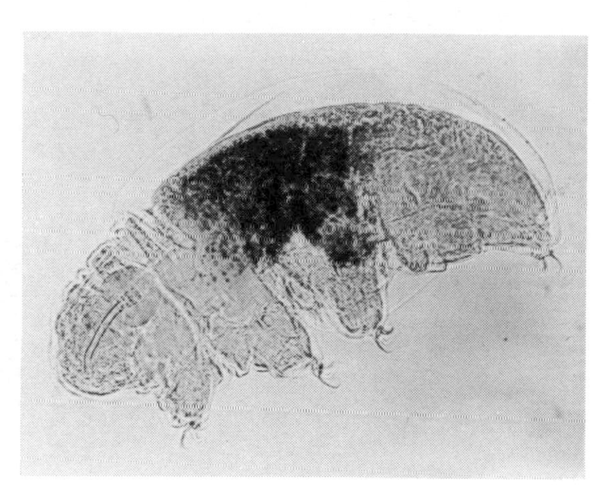

FIGURE 2-3

A water bear is a microscopic animal that often exists on mosses, liverworts, and lichens.

Photo courtesy: Richard Sayre

Since water bears generally inhabit mosses, this is one place where you can search to find them. Begin your search by obtaining portions of mosses from a wide variety of places: on trees, by streams, under trees, near trails, etc. Isolate each sample in a clean baby food jar or other suitable container; label its location. You should have at least fifteen samples before returning to the classroom.

Once in the classroom, carefully remove each moss sample from its container and gently wash the dirt from the base of the plants. Turn the moss cushion upside down in a small container. Add enough distilled water to cover only the green leafy part of the moss. Allow this to remain upside down from 24-72 hours; during this time the water bears and other moss life will leave the moss and eventually collect on the bottom of the container. (Refer to Sayre and Brunson, 1971.)

During the next few days use a medicine dropper to take samples of water from the bottom of the container. Examine under the microscope. If the tardigrades are not moving in the slide, you could be looking at a case skin or an inactive stage that may soon be active (Figure 2-4). Regardless, if you find one water bear there will more than likely be others, since thousands of water bears have been reported in just one gram of mosses.

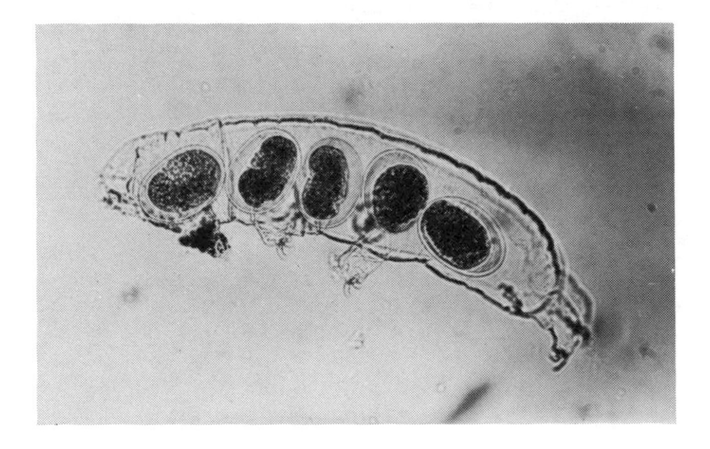

FIGURE 2-4

The eggs of a water bear may remain in the cast cuticle after the animal moults.

Photo courtesy: Richard Sayre

A better method that eliminates dirt problems and also provides a better chance of securing tardigrades involves searching specifically for trees mosses. Water bears abound in tree mosses, and since there is little or no dirt involved, the moss can be simply inverted in the container with little chance of dirt clouding the situation.

Another technique that may increase your chances of finding tardigrades is to gently "wring out" the mosses before taking water samples for microscopic examination.

Since water bears are well-adapted to temporary fresh-water living, they might be studied in terms of their habitat. You could gather mosses for such a study. Upon microscopic analysis a study should reveal at least three types of microorganisms: roundworms, rotifers, and water bears.

You could compare moss animals in tree mosses to those found in soil or stream locations, etc. This could be done in terms of numbers and types of organisms present.

In addition, there are a number of questions that could be asked:

1. What do water bears eat? (See Figure 2-5.)

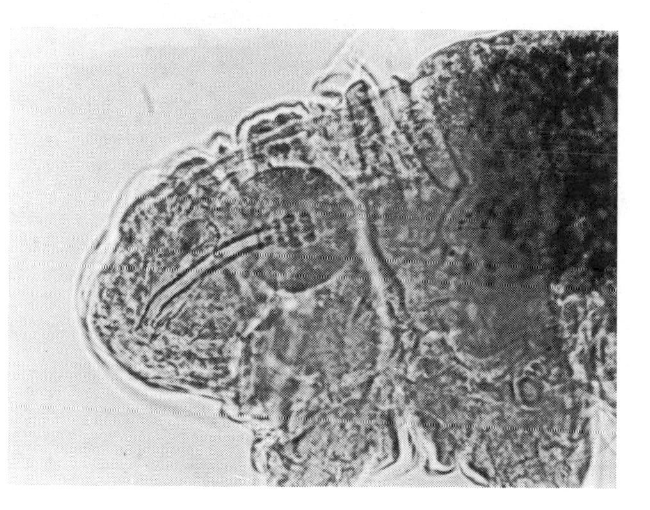

FIGURE 2-5

A water bear feeds with a muscular pump. Among its foods are roundworms, rotifers, and moss leaflets.

Photo courtesy: Richard Sayre

Isolate water bears for a few days and introduce three food sources: roundworms, moss leaflets, and rotifers. Do not offer some containers of water bears any of these foods. Make comparisons, in activity, size, and numbers, over a period of a week or longer. (Refer to Sayre, 1969.)

2. What happens if a water bear slowly dries out?

Allow a small container of water bears to lose water through evaporation. Record the number of days required for the total water to be evaporated. Observe the water bears during evaporation and record observations. You will more than likely observe them forming cysts or becoming inactive. (The water bears simply retract their heads and posterior ends and ball up into an inactive stage. See Figure 2-6.)

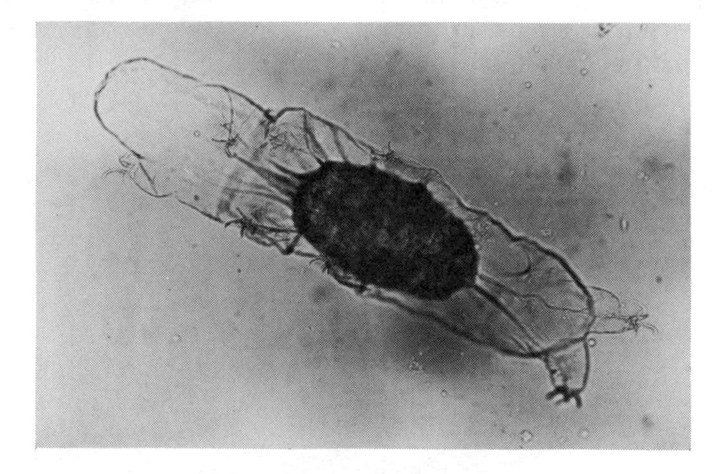

FIGURE 2-6

The inactive cyst of the water bear is seen as a dark, thick walled body within the original animal's cuticle.

Photo courtesy: Richard Sayre

---

3. Can the inactive stage of a water bear survive environmental punishments?

Wash the previously dried container with salt water, and if more than one container was used, place other containers under bright lights, inside the refrigerator freezer overnight or longer, inside the stove (or add boiling water to the container), or in other unfavorable conditions. When ready to revive the water bears, flood the containers with fresh water (distilled or tap water set out overnight).

4. How long does it take an inactive water bear to become active again?

You should find that water bears spring back to life in anywhere from four minutes to two days when water is returned to them. Examine the various containers with survivors and compare their activity. You may find that water bears are nearly indestructible. Crowe and Cooper (1971) report that over 90 percent of a water bear's body fluids can be removed without totally destroying it; Pennak (1953) reports that water bear's inactive stages have resisted brine solutions and both high and low temperature extremes for long durations of time.

If you have difficulty locating water bears in the field, you can always order them from Ward's (refer to listings at the end of the chapter). Actually there are only about 20 types of water bears in the United States, and since little work has been done with them in our country, your students may wish to broaden our knowledge of these unique organisms.

For further reference on water bears see LeGros (1958), Pamer (1965), and Riggin (1962) in addition to those authors previously cited.

While searching for water bears, or when examining various fresh-water habitats, you may discover rotifers. Rotifers have been studied for over 200 years, so although they may be a new discovery to your class, biologists have known about them for some time.

TYPES, CHARACTERISTICS AND HABITATS OF ROTIFERS

Much of the life we find in fresh water may have evolved from salt water, but rotifers are one of the few exceptions. They are thought to have evolved from fresh water and diversified to include over 1,500 species that may exist in mud puddles, ponds, lakes, or even in the film of water that covers moss plants. Rotifers are believed to exist world-wide; you can probably find them in your community in a variety of places, literally in any habitat that is moist with fresh water.

You can bring in mosses and, using the same methods as with water bears, you can usually, if not more often, find rotifers (Figures 2-7 and 2-8).

You can also go to a nearby lake or pond and scoop up matted leaves and vegetation along the shoreline. Place the debris in a large jar. Add enough muddy water to cover the bottom materials, filling over half the volume of the container. Allow the water to stand overnight. The next day, any free-swimming rotifers can be found near the edge of the surface film, with the greatest abundance on the side facing the light. A medicine dropper or pipette can be used to remove them for study. In addition, other rotifers can probably be found clinging to bottom debris.

If you have a plankton net, a plankton tow may reveal other rotifers. (Refer to Chapter Five for construction procedures for a crude plankton net.)

If time or facilities represent a problem, rotifers can be obtained from every major biological supply company. (Refer to listings at the end of the chapter.)

Once you have collected them, you may wish to culture rotifers for sustained use in the classroom. Although you may not wish to follow one popular technique–boiling manure in water, there are a number of alternative culture techniques at your disposal.

Rotifers have been cultured in solutions obtained from boiling 20 grains of powered rye in 100 ml. of water, one beef extract cube in 400 ml. of water, or eight to 30 flakes of dried breakfast oats in 100 ml. of water (Pennak, 1953). Other researchers have fed them *Euglena* or *Chlamydomonas,* or suggest dissolving 1 gm. of nonfat dried milk in a liter (quart) of distilled water (Lindner, Goldman and Ruzicka, 1961). Wards Culture Leaflet No. 11 recommends 0.1 percent malted milk powder boiled for one minute and allowed to stand for 24 hours.

One way you can introduce the rotifer is through a field-study approach. Attempt to find rotifers in a variety of fresh-water locations including holes of rotted trees, lake or pond debris, plankton tows, or in mosses or soil samples. Compare the types of rotifers in the different habitats. You should find an

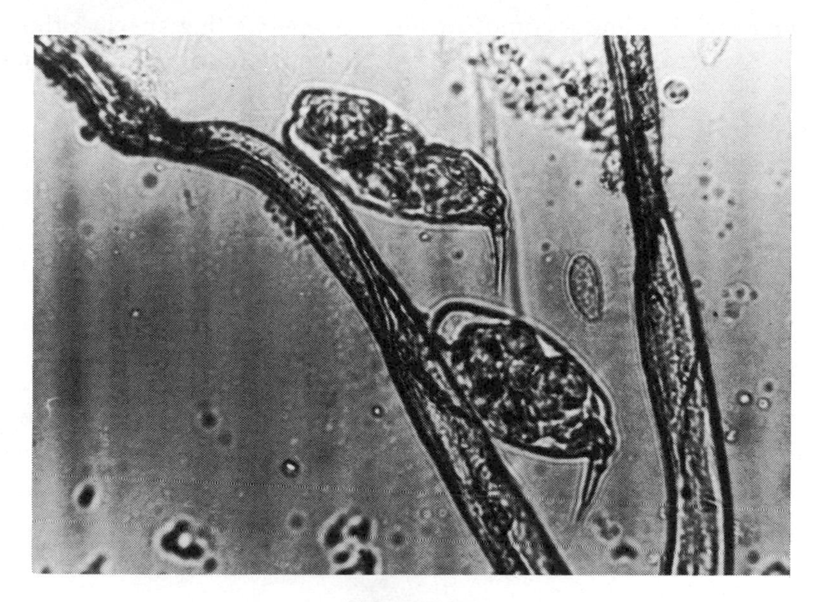

FIGURE 2-7

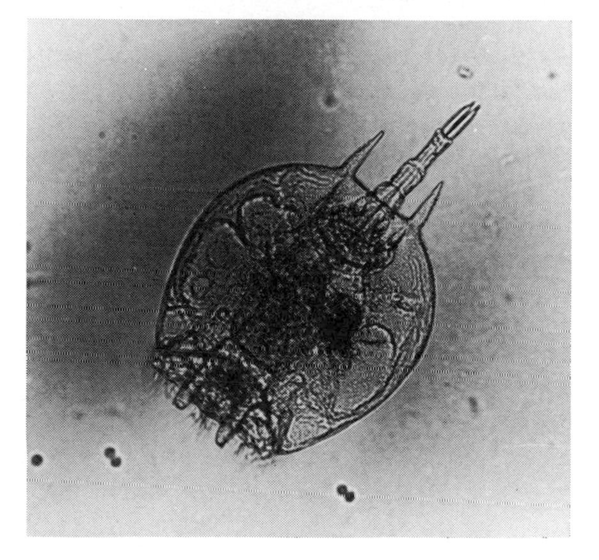

FIGURE 2-8

Two types of rotifers you may find in your cultures.

Photos by Stewart Harris; courtesy: Hampton Roads ETV Assoc.

inching or crawling locomotion in bottom-dwelling species; these use a cement gland to temporarily anchor them. On the other hand, you may find that some rotifers never touch the bottom, using border cilia and currents to keep them continually afloat as a part of the zooplankton in lakes. In general, rotifers may swim freely or utilize a crawling motion, depending on the specific habitat.

There are a number of questions that could be asked concerning rotifers:

1. Do rotifers have males and females in their populations?

Most rotifers are females; males are known only for a few definite species. The males often lack a digestive tract and always lack a mouth and anus. Their creation appears to be solely for reproductive purposes; the males swim about erratically in constant search of females throughout their two- or three-day, up-to-one-week existence.

2. How do rotifers reproduce?

You may discover that, for the most part, reproduction in rotifers is dependent on the females developing eggs which hatch without the presence of a male (parthenogenesis).

In the spring and fall you might search for males in the rotifer populations. Discuss the significance of the fertilized eggs; these can resist temperature extremes as well as dehydration. Normal unfertilized eggs cannot survive environmental hardships.

3. Do rotifers appear the same year-round?

You might initiate a year-long study to answer this question. Plankton tows from early May to July might specifically reveal physical changes in the rotifer, *Asplanchna priodonta,* which is markedly larger in July than in May. You might attempt to find out what causes the physical changes; this is an excellent area for serious science projects.

4. Can some rotifers survive when dehydrated?

Only a few types of rotifers can survive when dehydrated. This is usually restricted to those rotifers in a moss habitat or in similar temporary aquatic habitats. By allowing water to evaporate slowly from a container of moss-dwelling rotifers, and then allowing fresh water to return, you can observe first hand that some rotifers can form a resistant stage. The dried-out rotifers may survive as "dust" for years.

5. How do rotifers feed, and what do they eat?

Mix *Euglena* or *Chlamydomonas* with rotifers and observe the feeding action. Actually, rotifers can be identified by a feeding apparatus located in the upper portions of their bodies. The grinding "mastax," as it is called, is a complicated arrangement of muscles and jaws that can seize, tear or grind up food the rotifer brings in with its churning cilia (the cilia give a rotor appearance from which the rotifer receives its name). Offer them various organisms.

While searching for rotifers you might discover an animal that resembles a rotifer at a quick glance; however, a closer examination might introduce you to a gastrotrich, an animal that shares several similarities with rotifers.

A STUDY OF (FRESH-WATER) GASTROTRICHS, ANIMALS THAT ARE ALL FEMALES

Gastrotrichs are lesser known organisms that slightly resemble both flatworms and rotifers. In reality, gastrotrichs are considered to be a distinct group of animals that have a head region containing a few large tufts of cilia and a forked tail region containing two branches.

Although marine gastrotrichs generally have both sexes, there are no known males in fresh-water species. This means that every fresh-water gastrotrich can form eggs that will develop without being fertilized.

You can search for gastrotrichs in the bottom debris of bogs, puddles, marshes, and in the shallow margins of lakes filled in with decaying leaves. After taking a sample of mud and leaves, allow the debris to settle. Later remove a portion of the material at the boundary of mud and water and examine it under a microscope. The gastrotrichs can be seen gliding or swimming over pond debris, often displaying prominent spines or scales.

You can also search for gastrotrichs in sandy beach washings. However, if you wish to obtain pure cultures, or if you are unsuccessful in finding gastrotrichs, you can order them from biological supply houses (refer to listings at the end of the chapter).

To sustain a culture add two drops of uncooked egg yolk and some debris to 100 ml. of water. For pure cultures boil a 10 percent malted milk solution briefly before adding the powder. Allow it to stand for one week before introducing gastrotrichs.

If you have an older protozoan culture handy you may attempt to culture gastrotrichs here. Regardless, change the water in the upper third of the culture at least once weekly.

If you can discover a nearby habitat rich in gastrotrichs, you may not wish to sustain cultures since others can probably be obtained year-round. Peak collecting is usually in the summer and fall.

There are a number of questions involving gastrotrichs that you can consider:

1. How are gastrotrichs adapted to living on the pond bottom?

You may observe a number of bottom-dwelling adaptations, including ciliary movement, flattened body shape, a cement gland that offers temporary attachment, location of mouth, or the plates or scales and apines that sometimes armor the complete backside of the animal. While these spines offer protection, they appear to help some species to spring or leap through bottom debris.

2. Can gastrotrichs survive environmental hardships?

While female gastrotrichs in fresh water do not have a resistant male-fertilized egg as other creatures often have, the females can develop resistant eggs. *Slowly* dry out cultures, or *slowly* raise or lower temperatures to induce the production of resistant eggs. While a gastrotrich usually produces up to five eggs in its lifetime of less than a month, when the special resistant egg is formed it is generally the last egg of the animal.

3. How do gastrotrichs reproduce?

You can probably find several eggs developing in each gastrotrich. These appear as larger oval bodies; if patient, you may be able to watch the eggs greatly distort the gastrotrich's body as they leave the animal. The eggs turn hard on contact with water and possess many spines. The resistant egg can be distinguished from the typical egg by its larger size. In addition, the usual egg begins development shortly after departing from the parent.

While water bears, rotifers, and gastrotrichs have eggs that can resist unfavorable conditions, perhaps one of the best examples of resistant eggs can be found with brine shrimp.

THE IMPORTANCE AND SIMPLICITY OF RAISING BRINE SHRIMP IN THE CLASSROOM

Brine shrimp are one of the few fascinating creatures that can survive the salinity of the Great Salt Lake. This type of "fairy shrimp" is well-known for two characteristics: survival in salty habitats and production of a winter or resting egg that can survive various harsh environmental conditions. These eggs can exist for several years on the shelf of a classroom.

The resistant eggs make the brine shrimp *(Artemia)* valuable for classroom study. You can purchase viable eggs from most aquarium stores; they are also sold through most biological supply houses. The adults are also sold through local aquarium shops; eggs can be directly gathered from adult brine shrimp.

To initiate hatching, simply place a small amount of eggs in ocean water, or in an ocean (or sea water) "mix" that is also obtainable from supply houses or local aquarium shops. The saltier the water the better the chance for a greater hatching percentage. Also, if a higher concentration of sodium chloride is in the solution, the adults will reach a larger size and earlier sexual maturity.

After the eggs have been in the water for 24-48 hours, they will begin to hatch and young brine shrimp can be seen moving through the water. These young will develop through 14 different molts or skin-sheddings. They will reach maturity around the twelfth molt, or in about three weeks after hatching.

Brine shrimp can be sustained in the classroom for a life of up to nine months; they can be fed yeast and/or unicellular algae. However, for general purposes you will not need to feed brine shrimp since they can be studied temporarily and then fed to other invertebrates.

For example, you can investigate the hatching rates in various salinities, observe the hatching itself, or use the very young brine shrimp as food for *Hydra,* young fish, salamanders or other animals.

When using these salt-water (marine) animals as food for fresh-water life, capture them in a fine net, wash the net with fresh water several times, and then transfer them to the appropriate cultures. Another technique is to use a medicine dropper or pipette to transfer the young shrimp into a container of fresh water; by taking an immediate sample from the secondary container, very little salt will wind up in the fresh-water cultures. This secondary method is where you may wish to individually introduce brine shrimp to *Hydra,* etc; it also provides an alternative if you do not have a fine net at your disposal.

When making up salt solutions for brine shrimp, *do not* use iodized salt such as you find on the dinner table. Salt mixes can be purchased from aquarium shops and although they are relatively expensive (a few dollars), enough mix is available in one five-gallon package to hatch brine shrimp all year long.

Since the eggs can be stored on the shelf for several years, you can easily set up the conditions for studying brine shrimp on any given day. In addition, the costs are minimal considering that the organism itself may be studied or used as food for other living organisms in the classroom. Costs are greatly minimized if you live near the ocean since a suitable habitat for hatching can be easily obtained in one scoop of water. A few dollars can purchase an almost infinite amount of brine shrimp eggs.

While the brine shrimp *(Artemia)* provides both a suitable food organism and a subject for individual consideration, its cousin, the water flea *(Daphnia)* also fits into these categories, but it normally thrives in fresh water.

OBSERVING A COMPOUND EYE AND CRUSTACEAN HEART IN DAPHNIA

Water fleas, or *Daphnia,* have much to offer. While they provide excellent food for *Hydra,* fish, dragonfly larvae, and other pond organisms, their transparency offers a view into the workings of a compound eye and crustacian heart.

You can collect *Daphnia* in ponds or lakes throughout the year although there are one or two peak population periods during the year. If you discover when water fleas hit their "pulses" in your nearby ponds, you will know when to best expect them in future years.

A plankton tow will often provide *Daphnia* in large numbers, but if you take large water samples along the shorelines, you may find *Daphnia* swimming in the water. In any case, you can easily obtain water fleas from biological supply houses.

Daphnia can be cultured in a large jar or aquarium of pond water. Avoid using tap water directly since *Daphnia*, like most fresh-water life, may be killed by the dissolved gases; if you wish to use tap water, allow it to stand at least 24 to 48 hours prior to use. Since some tap water may contain (local) harmful substances, some authorities suggest using only distilled or actual pond water.

Boil an egg and mix its yolk with about one quart of distilled or pond water. This should be added to the container of *Daphnia* until the water becomes faintly cloudy. Do not add too much; remember that more food can always be added later. After several days the water may become clear, at which point you can add more of the hard-boiled yolk and water suspension. (Brewer's yeast or cake yeast may be substituted.)

Although the bottom of the *Daphnia* container may fill in with debris and sediment, do not remove this material since *Daphnia* eggs may be present. When more food is added, many of these eggs will hatch and add to the population. If the population becomes too dense, remove some of the numbers. Overcrowding can endanger the entire culture.

Keep the water fleas at room temperature, occasionally break the surface skim that usually forms, and replace any evaporated water with fresh pond water.

When investigating *Daphnia*, you should definitely consider the compound eye. Questions concerning the eye can include:

1. How many lenses make up the *Daphnia's* eyes? (Few to numerous.)
2. What controls the eye? (Three pairs of muscles.)
3. Is the *Daphnia's* eye closer in appearance to an insect's eye or a human eye?
4. Does the *Daphnia's* eye respond to light, an approaching object or change in temperature?

The water flea's heart, which is located on its upper back, offers an excellent view of a crustacean heart. A number of questions can be asked concerning the heart, including:

1. How many heartbeats per minute occur at room temperature? (Take the number of heartbeats in ten seconds and multiply by six.)
2. How many heartbeats occur per minute if a small piece of ice is added next to a water flea?
3. Does the presence of adrenalin on the slide affect the heart rate?
4. If slightly heated water is introduced to a water flea, is the heart rate affected?

In addition to these simple studies, you can observe the brood chamber near the heart and often find eggs present. You may even find several large opaque eggs that are the product of sexual reproduction. (*Daphnia* are normally females that reproduce without the presence of males.) The special opaque resting eggs can withstand drying and freezing conditions. You might test their resistence by placing these eggs in the freezer or under a light bulb for several days or longer.

Another interesting study can involve cyclomorphosis. This occurs when the *Daphnia's* head becomes elongated during the course of a year. Often in midsummer the heads of certain water fleas look like helmets or hoods. Scientists are not absolutely certain what causes this change in appearance. This area is excellent for initiating student projects; however, you might have to encourage summer projects in this area since most of the significant changes take place during the summer. (Duplicating exact summer conditions in the classroom is *most* difficult.)

Once you have obtained and/or cultured brine shrimp and *Daphnia,* you are now ready to secure predators to help control their numbers. One predator that feeds on both brine shrimp and water fleas is the *Hydra.* As a matter of fact, a *Hydra* will even swallow a blood clot.

WHY A HYDRA WILL SWALLOW A BLOOD CLOT

You can find the *Hydra* in many fresh-water ponds. (See Figure 2-9.) Its long slender body, usually topped with a distinct mouth and five to six arms, indicate its way of life: by using numerous stinging cells on its arms, the *Hydra* kills and ingests water fleas, small insects, worms, and even young fish.

FIGURE 2-9

***Hydras* are valuable invertebrates that can be useful when investigating growth, ingestion, reproduction, regeneration, and reassociation.**

Photo courtesy: Turtox

You can test the *Hydra's* stinging cells by placing a water flea with a *Hydra* on a glass slide. Observation may show how *Hydra* victims are killed almost instantly upon contact with the arms and stinging cells. (Larger victims may resist for a few minutes and some may even escape, but the variety of stinging cells usually prohibits escape—certain stinging cells puncture and paralyze, others entangle or adhere to the victim.)

Research has shown that the *Hydra* feeds on response to a certain chemical, which explains why it will not sting debris that encounters its arms. The chemical is one that is found in living things; it is known as glutathione. Glutathione is found in yeast, shrimp, and even your own blood. Because your own blood contains this chemical, *Hydra* will attempt to swallow a blood clot that is introduced into their culture (Loomis, 1955).

You may wish to experiment with this chemical. You can mash up live brine shrimp or *Daphnia* and smear the juices on bits of cotton, introduce the cotton to *Hydra* and record responses.

If you filter a yeast solution you can concentrate glutathione. *Hydra* have even been observed turning inside out trying to "eat" a concentrated yeast solution.

Discussion of glutathione provides interesting commentary on whether man can be made to prefer certain foods treated with a similar chemical.

Hydra can be used in a number of ways in the classroom, as you will find out shortly. Therefore you might be interested in collecting and maintaining an ample supply of *Hydra.*

You can collect *Hydra* on leaves found in ponds. They can be picked directly from the leaves, or the water can be agitated until the tiny *Hydra* are released from their attachment points and float about. You can remove them with a large pipette. If *Hydra* are scarce locally, they can be ordered year-round from biological supply houses. (Refer to listings at the end of the chapter.)

Hydra can be cultured in the classroom by keeping them in freshly changed pond water under moderate light. With less than a dozen *Hydra* in a large aquarium or finger bowl, feed each one at least one or two *Daphnia* daily.

If you do not wish to feed the *Hydra* individually, place a few droppers full of water fleas in the container each day. However, be sure to remove any debris that remains after a feeding session, and change at least the water in the upper third of the culture *daily.*

Daphnia can be replaced or supplemented with brine shrimp, but be especially careful to wash the brine shrimp thoroughly before introducing them to the *Hydra* culture. As with the water fleas, be sure to remove any debris that remains after the feeding sessions.

SELECTED REGENERATION EXPERIMENTS WITH HYDRA, PLANARIA, SPONGES, AND TUBIFEX WORMS

While *Hydra* have an interesting ingestion process, they also have a remarkable ability to grow back or regenerate lost body parts. Regeneration is found in all animals to some degree, but while man can at best replace liver or skin cells, others such as the *Hydra* can survive if cut into unrecognizable pieces.

Hydra

Hydra not only regenerate lost body parts, but they can reassemble if the cut portions are placed near each other (reassociation).

You can experiment by cutting off tentacles or part of a bud, or cutting an entire *Hydra* in half. You might determine how many sections will grow back into complete *Hydra,* or if certain areas do not regenerate.

You can demonstrate reassociation or reconstitution by cutting up five or six *Hydras* and placing the portions in a depression slide or very close together in a small container. The parts should begin to fuse together within one week. You can vary this by placing only tentacles or head regions together. Record the time involved for at least one individual to form from the pieces.

In addition to regeneration and reconstitution, you may wish to graft part of one *Hydra* to another. In order to graft *Hydras,* the minimum supplies you will need include a sharp razor or dissecting needle and a dissecting microscope. You can try placing two heads on a single body, or attempt any number of combinations.

Like the *Hydra,* planaria also make excellent specimens for regeneration experiments.

Planaria

Planaria are flatworms that can be found in many ponds that are filled with decaying leaves and debris. They can be collected by placing shoreline materials in a large container and covering with a few additional inches of water. Allow this material to settle overnight and by the following morning planaria can often be found crawling up the sides of the container.

If you can purchase a plastic medicine dropper and cut back the opening somewhat, you can use this to remove planaria to separate culture dishes. Since planaria may hold on tightly to the sides of the container, wash them back and forth with water until they loosen. Do not place too many planaria in one container. You will soon find out how many planaria a culture dish will support.

Use spring, pond, or distilled water in your cultures and avoid the use of tap water. Keep it in a shaded place and add a few objects for the worms to hide under.

If you are seeking planaria from streams it may be best to "bait" them. Place a small portion of beef liver in the water to attract them. After fifteen minutes to half an hour, remove the liver and shake off or manually remove any worms on the meat.

Planaria can be ordered from most biological supply companies. You are offered choices of brown, white, or black planaria.

Planaria can be fed many foods in the classroom. Pieces of earthworms, fresh beef liver, crushed snails, or the yolk of a hard-boiled egg offer suitable foods for many planaria. The brown planaria seem to prefer hard-boiled egg yolks and a small amount can be added once or twice weekly to their cultures. After the food has been added, allow half an hour for them to feed; afterwards remove the planaria to a clean container of pond water. Do not overfeed; keep in mind that planaria can live from three to twelve weeks without feeding, although they will continually grow smaller. Too much food in a culture, without a replenished supply of water after feeding, could easily kill them.

Regeneration of Planaria

For best regeneration results, use the brown planarian *(Dugesia tigrina)* or the white planarian *(Phagocata oregonensis)*.

Allow the worm to stretch out on a piece of glass, a cube of ice, or the flat side of a cork stopper. Use a razor blade to make the incisions. You might experiment by cutting planaria directly behind the head, in several sections, or slicing the head in several sections. You may discover that the further back the incision is made from the head region, the less frequently the regenerated worm forms a complete head.

After cutting the worm into pieces of varying shapes and forms, place the sections in fresh pond water and store in a cool, shaded place. Do *not* feed while regeneration is taking place.

If the head is sliced repeatedly, and the portions are slightly separated, you may grow a multi-headed planarian (Figure 2-10); however, if the portions touch, they may grow back together as in the cases of incomplete incisions.

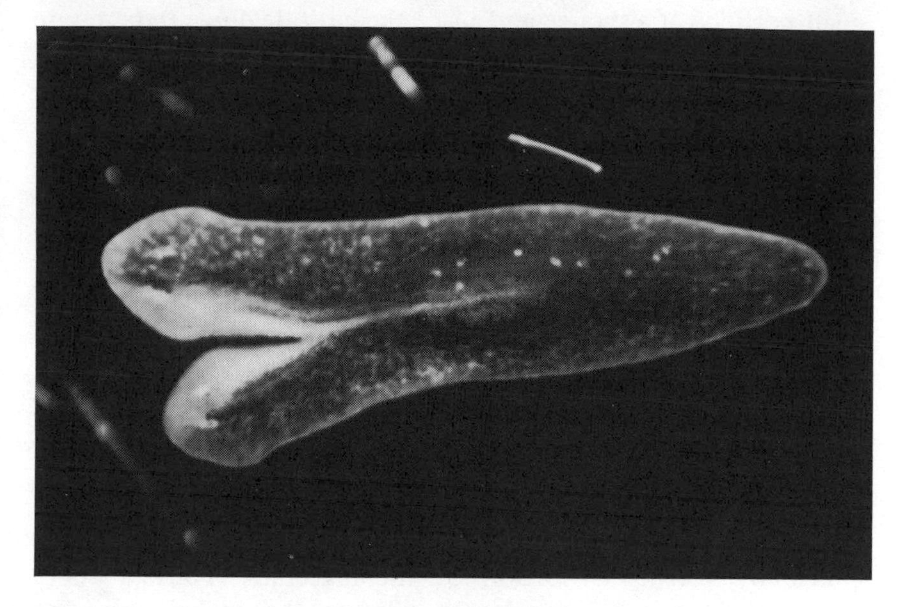

FIGURE 2-10

If you carefully slice the head of a planarian, you may be able to grow a two-headed planarian like this one.

Photo courtesy: Carolina Biological Supply Company.

Since planaria may accept portions of other planaria, you may also wish to attempt some grafts.

Another creature that regenerates well is the sponge.

Sponges

Fresh-water sponges are found in ponds, streams, lakes, rivers; they grow in a matlike fashion over rocks, logs and twigs. Sponges can be gathered in shallow water without any special equipment; a long-handled rake may help in securing them from deeper waters. Regardless, use caution when removing samples.

It is doubtful that you will be able to keep sponges alive in the classroom for longer than a few months. For best results, use small portions of a sponge and place the cut side down (where you removed the sponge); allow this side to grow onto a glass slide. Keep the slide in a shallow aquarium, preferably where other life forms from the natural environment are mixed. Providing a gentle water circulation may encourage their survival.

Fresh-water sponges *(Spongilla)* can be supplied by Carolina Biological and Connecticut Valley supply houses; refer to listings at the end of the chapter.

Regeneration with Sponges

Sponges have remarkable powers of regeneration—not only can they replace missing parts, but if strained through fine bolting silk after being completely macerated, the resulting parts can reassemble. After several days the portions can form a new sponge.

You may attempt to macerate the fresh-water sponge for several purposes. You can search for the specialized cells of the sponge by placing a portion of the macerated body under a microscope, or you can simply pile up the majority of the macerated portions in a clean container of water for reassociation. As with planaria and *Hydra,* do not feed during the reassembling period. Record observations.

Tubifex Worms

Tubifex worms make excellent specimens for regeneration experiments, but you have to work with a small and possibly active invertebrate. However, you can slice sections off the head region, or elsewhere, and observe regeneration.

Collecting and culturing techniques for *Tubifex* are included in Chapter Eight.

LOOKING INTO SEGMENTED WORMS

Tubifex worms not only demonstrate regeneration, but they also, as do many of the aquatic segmented worms, afford an opportunity to view life processes in action. You can secure aquatic segmented worms in pond debris or from biological supply houses.

Beneath the microscope these worms can reveal peristalsis or the movement of intestinal waves, the bristles or setae, and the blood flowing in veins. Once you slow down one of these worms, the action of muscles and internal processes make dissections seem like primitive ways to study life.

Why not instead observe life in action by obtaining these abundant creatures that summarize invertebrates beneath the microscope: fascinating, complex, and mysterious organisms, organisms that bring life processes directly to the viewer.

REFERENCES

Carolina Biological Supply Company Culture Sheets, Burlington, North Carolina.

Crowe, John H. and Alan F. Cooper. "Cryptobiosis," *Scientific American,* vol. 225, no. 6 (1971), pp. 30-36.

Galtsoff, Paul S. *et al. Culture Methods of Invertebrate Animals.* Ithaca, New York: Comstock Publishing Company, 1937.

LeGros, A.E. "How to Begin the Study of Tardigrades," *Countryside,* vol. 18 (1958), pp. 322-330.

Lindner, L., H. Goldman and P. Ruzicka. "Simple Methods for Rotifer Culture." *Turtox News,* vol. 39, no. 74 (1961).

Loomis, W.F. "Glutathione Control of the Specific Feeding Reactions of Hydra," *Ann. N.Y. Acad. Science,* vol. 62 (1955), pp. 209-228.

Nichols, David, John Cooke, and Derek Whiteley. *The Oxford Book of Invertebrates.* New York: Oxford University Press, 1971.

Pennak, R.W. *Fresh-Water Invertebrates of the United States.* New York: Ronald Press Company, 1953.

Pramer, D. *Life in the Soil: High School Biology–A Laboratory Block.* Boston: D.C. Heath and Company, 1965.

Reid, George K. *Pond Life, A Golden Nature Guide.* New York: Golden Press, 1967.

Riggin, G.T., Jr. "Tardigrade of Southwest Virginia," Technical Bulletin 152, Virginia Agricultural Experiment Station, Blacksburg, Va.

Sayre, R.M. "A Method for Culturing a Predaceous Tardigrade on the Nematode *Panagrellus Redivivus,*" *Trans. Amer. Microsc. Soc.,* vol. 88, no. 2 (1969), pp. 266-274.

Sayre, R.M. and Linda K. Brunson. "Microfauna of Moss Habitats," *American Biology Teacher,* vol. 33, no. 2 (1971), pp. 100-102.

Turtox Service Leaflet No. 39, *"The Fresh-Water Hydras."* Chicago: CCM: General Biological, Inc.

Ward, Henry B. and George Whipple. *Fresh Water Biology.* New York: John Wiley and Sons, 1959.

Ward's Culture Leaflets: No. 7, "Rearing *Hydra* in the Laboratory"; No. 8, "Culture of Planaria and Other Worms"; No. 11, "Culture of Rotifers and Other Invertebrates"; No. 12, "Rearing of Daphnia and Related Arthropods." Ward's Natural Science Establishment, Inc., Rochester, N.Y.
Welch, Paul S. *Limnology.* New York: McGraw-Hill Book Company, 1952.

AUDIO-VISUALS*

Films:

Brine Shrimp (Wards)
Flatworms (Platyhelminthes) (EBEC)

Filmstrips:

Classification of the Flatworms (McGraw-Hill)
Roundworms and Some Minor Phyla of Animals (McGraw-Hill)
Zoology: Sets 1, 2 (Invertebrates) (McGraw-Hill)

Film loops:

A Loricate Rotifer (Holt)
An Illoricate Rotifer (Holt)
Animal Reproduction (10 loops) (Ealing)
Cladoceran (Thorne)
Copepod (Thorne)
Feeding Behavior of Hydra (BSCS)
Hydra (Thorne)
Leech (Thorne)
Planaria (Thorne)
Planarian Behavior (BSCS)
Rotifer (Thorne)
The Lower Invertebrates (6 loops) (Holt)
The Invertebrates (25 loops) (Ealing)

*Audio-visuals generally intended for upper elementary through high school audience; refer to glossary (page 211) for key to abbreviations and addresses of companies.

LIVING INVERTEBRATES: FOR CLASS OF 12 OR MINIMUM PRICE*

	Bico	*Carolina*	*Conn.*	*CCM-Turtox*	*So. Bio.*	*Stansi*	*Schettle (Moguled)*	*Stein. (Nasco)*	*Wards*
Water bears	–	–	–	–	–	–	–	–	3.50
Rotifers	1.50	1.25	1.25	1.85	1.75	1.25	1.00	2.25	2.50
Gastrotrichs	–	1.75	1.50	–	2.20	1.75	–	–	2.50
Brine shrimp (eggs)	1.50	1.75	2.50	2.00	1.70	1.75	1.50	1.50	4.00
Daphnia	1.50	1.50	1.50	4.00	2.00	2.00	1.25	2.25	2.75
Hydra	2.25	2.50	2.35	2.60	2.40	2.50	2.25	3.50	3.25
Planaria	2.50	3.00	2.75	3.60	2.60	3.00	3.00	3.00	3.75
Sponges	–	6.50	4.00	–	–	–	–	–	–
Tubifex worms	3.25	2.25	–	6.00	–	–	–	2.25	–
Aquatic Segmented worms: Aelosoma, Dero, or Stylaria	–	2.25	1.75	2.35	–	–	–	–	–

Higher prices usually indicate larger amounts. The prices were in effect at the time this book was written. It is suggested that when you write for cultures, you request the organization's current price list.

While not an official biological supply house, Edmund Scientific Company, Edscorp Building, Barrington, N.J. 08007, offers the following for a class of 25: Planaria, $5.00; *Hydra,* $4.50; and *Daphnia,* $3.50.

*Refer to glossary (page 210) for key to abbreviations and addresses.

THREE

Insects Around Us

With three pair of legs, compound eyes, and the ability to fly, the typical insect is one of 750,000 kinds that surround man. Yet, how many living insects did you have in your classroom last year? Perhaps we neglect them because they are most abundant during the summer, but many species are available year-round; others may be actively collected in the spring and fall. (See Figure 3-1.)

Once you have obtained a living adult insect you will need a method to limit movement during observation or experimentation sessions. One way to immobilize an adult insect is through the use of carbon dioxide, the same gas that humans exhale.

HOW TO SAFELY "KNOCK OUT" AN INSECT

Since an insect breathes through small openings in its body, since it lacks the large vertebrate brain, and since it is often extremely small and delicate, an insect must be handled differently from larger laboratory specimens.

Use a fine net to capture the adult insect and carefully transfer it to a jar; any size jar may work, but a smaller container is better than a larger one. Punch a few air holes in the container top. Make one air hole large enough to accept a rubber hose that will be used later. Once you have made the larger hole, cover it with masking tape.

Now that the container is ready, you will need to construct a gas-generating apparatus. A 500 ml. or 1000 ml. flask can be used to hold an ample supply of marble chips (calcium carbonate or

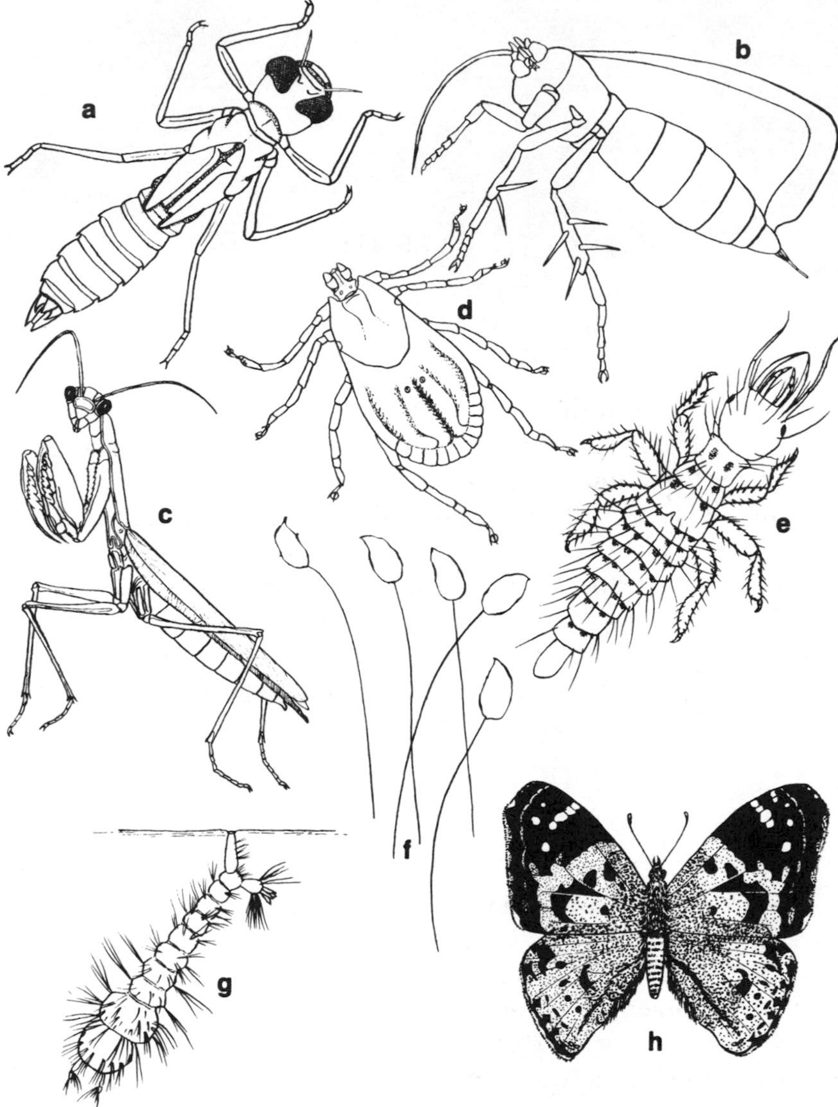

FIGURE 3-1

Chapter 3 is concerned with insects and a few additional types of arthropods including these: (a) dragonfly larva (nymph), (b) wax moth, (c) praying mantis, (d) tick, (e) lacewing larva, (f) lacewing eggs, (g) mosquito larva, (h) painted lady butterfly.

Original drawings by Sheralyn Lerner; after various authors; not to scale limestone). Use a two-hole stopper with one hole filled with a thistle tube; the other hole should occupy a glass tube connected to a rubber hose. (See Figure 3-2.)

Prepare a diluted solution of hydrochloric acid; a 10 percent solution is recommended but not absolutely necessary. Be extremely cautious when handling hydrochloric acid; immediately wash any spillage with large amounts of water, wear safety glasses, and remember *not* to pour water into a container of acid.

When ready to immobilize the insect, pour a small amount of acid into the thistle tube of the gas-generating apparatus. Place and hold the rubber hose over the special opening in the top of the insect jar. Continue to gently add acid until the insect is "knocked out." Do not move the container during the addition of the gas; allow the heavier-than-air carbon dioxide to settle on the bottom of the container. It may take several minutes for the insect to become totally immobilized.

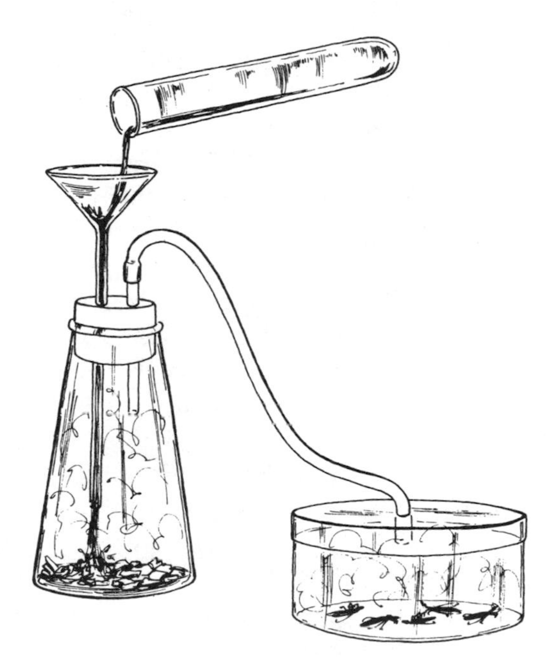

FIGURE 3-2

When dilute hydrochloric acid is added to marble chips (calcium carbonate), it creates carbon dioxide. This gas provides one way of immobilizing insects.

Once immobile, the insect can be quickly removed to a study chamber or used for whatever purpose is intended. The insect may not revive for a minute or two; at first it may appear sluggish.

While on an open surface, the insect can be kept immobile by allowing the gas to flow directly over it. However, do not deprive the insect of oxygen for too long.

When dealing with fruit flies and certain other insects you can also immobilize them with ether, but avoid prolonged exposure to the gas. A few minutes in the freezer may also immobilize insects.

In some instances you may wish the insect to remain completely "unconscious" or immobile for long periods of time. This is especially true when observing "blood" moving through the wing of an insect.

TECHNIQUES FOR OBSERVING THE BLOOD IN THE WING OF A LIVING INSECT

While it is difficult to observe blood flowing through many larger organisms, the thin wings of a living insect offer direct windows to the circulatory system.

First, capture a living damselfly, dragonfly, or large cockroach (Figure 3-3). The larger the specimen the better. (Cockroaches can be ordered year-round from biological supply companies; damselflies and dragonflies can be captured in fine nets in early fall and spring.) Place the specimen in a jar that can be used to immobilize the insect with carbon dioxide.

Obtain a paraffin block (sealing wax from a grocery store will suffice). Hollow a small chamber for the insect and melt part of another block in a small pan.

Once the chamber has been prepared, immobilize the insect with carbon dioxide and transfer it to the chamber on the wax block. Lift up one wing (both large, thick outer covering, and more delicate wing portions) and slowly pour melted paraffin over the balance of the insect, especially the legs. This paraffin will not be hot enough to injure the insect, but when it hardens the insect will be immobilized. If done quickly the insect should not awaken until the paraffin solidifies.

Cut away the excess paraffin but be careful not to remove too much. Hold the free delicate wing of the insect under the

FIGURE 3-3

Large cockroaches make excellent insects to use when observing the blood flowing through the wing of an insect.

Photo by Stewart Harris; courtesy: Hampton Roads ETV Assoc.

---

microscope and observe the thinnest sections of the wing closest to the body. Situating the wing under the microscope may be difficult and observing the blood may require throwing the veins and tissues slightly out of focus. Use medium and highest powered objectives for best viewing.

The blood cells of the insect destroy invading bacteria and block up tears in the body tissues when the insect is injured. Students should observe several shapes of blood cells. You might point out that the insect's blood does not carry oxygen as our blood does, and that an insect's circulatory system is a combination of open areas and veins. It does not have a closed system like ours. You may also observe a change in direction of blood flow under the microscope.

When completed, the cockroach (or substitute) must usually be destroyed since removing the wax often breaks off its legs. There are more favorable endings of a study when the insect can be released alive, back to where it came from. Thus, the dragonfly larvae offers an interesting insect for study; it can be maintained in the laboratory for months and then be safely returned to its habitat.

HOW TO SECURE AND SUSTAIN DRAGONFLY LARVAE IN THE CLASSROOM

From late spring until early fall, dragonflies mysteriously fill the air, only to disappear during the colder months. To many teachers this means that dragonflies are unavailable to the classroom during their apparent absence from the environment.

Yet, within the lakes and ponds, young dragonflies exist year-round as developing larvae or nymphs; in some cases a dragonfly may spend several years in the lake as a larvae and only a month or so in the air as an adult. So we need to think of the dragonfly more as a "water creature" than an "air creature."

The young dragonflies are ideally suited for classroom study. They are usually abundant in surrounding lakes and ponds; they are easy to obtain; and they are easy to maintain in the classroom.

To secure dragonfly larvae from their habitat, first purchase a white photographic tray, and either make or purchase a fine dip net. A fine screen may substitute or assist the dip net. Obtain a plastic eye dropper and cut off the end to make a wider opening.

Locate a nearby pond, stream, or lake, preferably one that has been seen with adult insects around it in the summer. Fill the white photographic tray with about an inch of water and place it on level ground. Use the dip net to scoop up leaves and sand from the lake margins. Repeatedly dip the net full of debris in clear water before emptying it in the photographic tray.

Carefully remove debris, shaking the leaves in the water before removing them. Observe the tray for several minutes, stirring the water gently at periodic intervals. Remove dragonfly larvae with a wide medicine dropper and place in small baby food jars about one-third full of pond water. Refer to diagrams, Pennak (1953) and Ward and Whipple (1959) for appearances of dragonfly larvae.

When in a stream environment, place the fine screen downstream and lift up rocks and stir mud above it. After a minute or two shake the screen over the white photographic tray. Transfer to small baby food jars filled partially with water.

If securing the larvae from their natural habitat is impossible, dragonfly larvae can be secured from biological supply houses, although they are sometimes listed as scarce in the winter.

For sustaining them in the classroom, use large fingerbowls or other containers that allow a large volume of water with a large surface area. Keep the water shallow to maintain the large surface area per volume of water; this will allow oxygen to diffuse adequately in the water so that additional plants or mechanical pumps are not necessary. However, you might add a few plants to assist oxygenation of the water; these may also provide objects on which some dragonflies cling. If the dragonfly was found in the sandy lake bottom, place sand from the natural habitat on the bottom of its container.

Dragonfly larvae do not require feeding in the natural habitat during winter. If kept in cool temperatures under normal day-length lighting, dragonfly larvae may require little feeding in the classroom. However, periodic feedings of *live Daphnia* (water fleas), aquatic earthworms, or brine shrimp (washed before adding to culture) make suitable prey for this voracious predator. Be sure to place only one or two larvae together, and change the water occasionally.

To study dragonflies, there are a number of questions that might be asked:

1. How many legs does the larva have? Why is the number significant?

The three pair of legs indicate its relatedness to other insects. Students could discover this if given a glimpse at a number of insects mixed among other organisms.

2. Does the larva have wings?

Note the protective wing pads on the back of the specimen. The wings will unfold when the larva leaves the water to become an adult.

3. How does the dragonfly larva breathe? How does the adult dragonfly breathe?

You should discover the pulsating motion at the rear of the dragonfly larvae. There are gills within the rear cavity and the water is repeatedly washed over the gills by the muscular action. On the other hand the adult breathes through openings in its body. The oxygen is able to penetrate deep within the adult's body and diffuses through the nearby cells.

4. How does the dragonfly larva feed? How does the adult feed?

You can discover a fascinating feeding sequence when placing *Daphnia* with the dragonfly larvae. The larva has a spoon-shaped feeding apparatus that extends approximately one-third its body length to snare victims. The adult, on the other hand, has much less prominent jaws that snare tiny insects while on the wing. This might be a good time to dispel the notion that dragonflies attack people; dragonflies fly near houses at sunset in their search for food, small swarms of smaller insects. Their feeding apparatus are their feet, which form a basket apparatus to capture the prey, and their jaws, which are designed to eat smaller organisms. So, if the dragonflies move toward a lighted area where people stand, it's only because they seek the insects that are attracted to the lights. There could be no purpose in attacking people.

5. How does the dragonfly larva grow?

Students should be able to observe a series of moltings in the laboratory if the larvae are fed regularly. The dragonfly larva does not pass into a pupa stage as do some insects. After it reaches a certain size it crawls out of the water, splits its last larval skin and emerges as an adult. It may rest for 24 hours before taking to the air. However, laboratory-raised specimens rarely emerge as adults, so it might be advisable to release specimens back into the native waters after studying them. This should increase their chance of emerging as adults in the spring or summer.

6. How do dragonflies suddenly appear in large numbers "all at once" in the spring or summer?

The dragonfly larvae experience *diapause,* growth stoppage triggered by a response to change in day length. The faster growing larvae enter diapause early in the fall. In the spring the smaller specimens leave diapause earlier and continue to grow while the larger specimens wait in their developmental arrest condition. The final result is that the majority of the species emerge as adults at approximately the same time. Since the adults must mate and lay eggs in a relatively short time, this process helps insure survival.

7. What is the value of a dragonfly?

While in the lake, the dragonfly larva is a voracious feeder; it helps control the numbers of other small species as well as provide food for larger organisms. When in the air the dragonfly is a predator of smaller insects, some of which bother man. Yet the dragonfly has another value. It is a living reminder that insects have survived for millions of years; the dragonfly is a living fossil. So, we don't need to justify the existence of this creature–it was here long before man; besides, the beauty of this insect alone justifies its existence, if justification is necessary. For further reference see Pennak (1953) and Ward and Whipple (1959).

When searching for the dragonfly you may discover a host of other invertebrates (Figure 3-4). You may discover the midge larva, usually red in color and abundant in the bottom mud. You may discover the damselfly larva that is similar to the dragonfly larva except for a three-pronged tail and a somewhat larger head region in comparison to body width. Others that are common in many ponds and streams are: stonefly larvae (do not thrive well in the classroom), mayfly larvae (three-pronged tail and vibrating gill plates–do not thrive in polluted water), whirligig beetles, water boatmen, back swimmers, water scorpions, water striders, water mites, crayfish, diving beetles, *cyclops,* amphipods (sideswimmers), isopods, and many others.

For further reference on aquatic life refer to Borror and Delong (1964), Gorvett (1970), Geltsoff (1937), Needham and Needham (1962), Reid (1967), Pennak (1953), and Ward and Whipple (1959).

Aquatic invertebrates are excellent organisms to study in the field and in the classroom. They are not only abundant, but as in the case of the dragonfly larvae, they are extremely easy to keep in the confinements of the classroom, requiring little daily care.

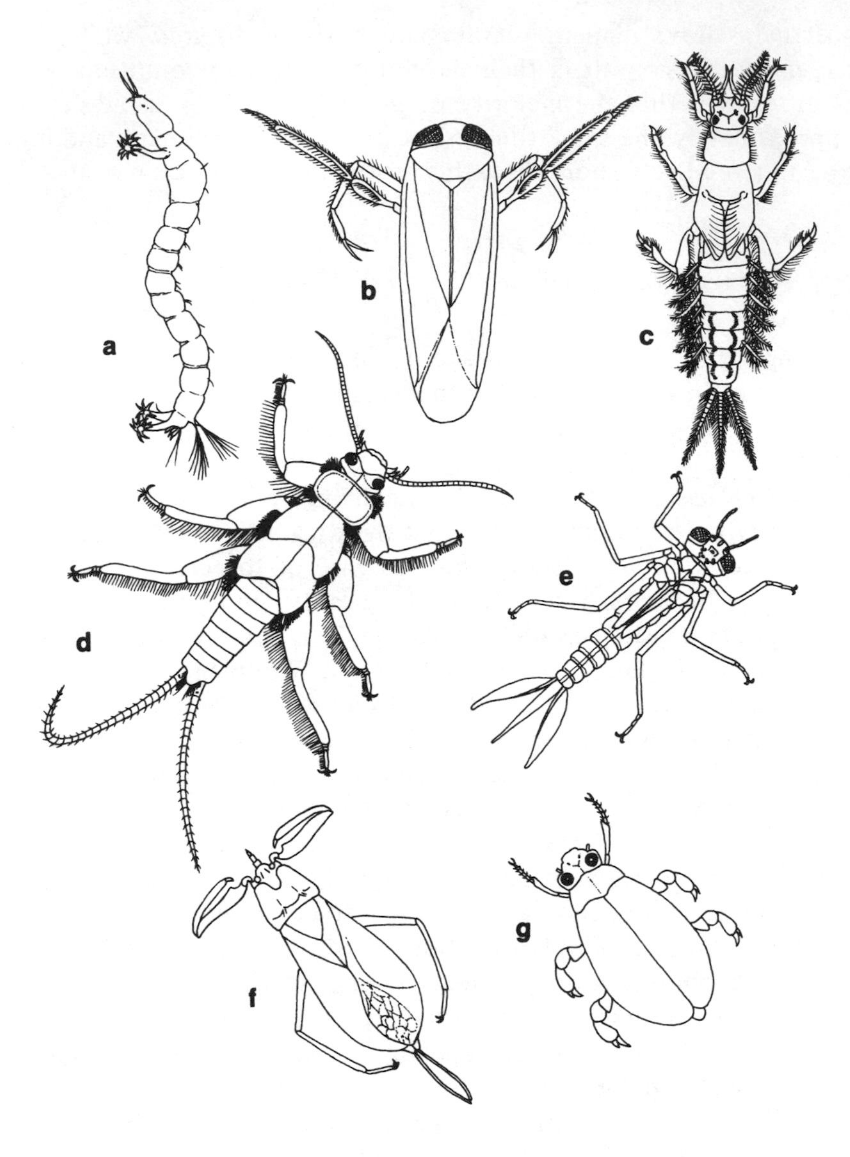

FIGURE 3-4

Other aquatic insects include the following:
(a) bloodworm (midge larva), (b) water boatman, (c) mayfly larva, (d) stone fly larva, (e) damselfly larva, (f) water scorpion, (g) whirligig, adult.

Original drawings by Sheralyn Lerner; after various authors, not to scale.

However, classroom observation is largely limited to the larval stages of the life cycle for many of the insects. Yet you can easily observe the entire life cycle of the greater wax moth directly in the classroom.

RAISING THE WAX MOTH: LITTLE TROUBLE, LARGE REWARDS

The greater wax moth, *Galleria Mellonella,* is an insect that parasitizes bee hives. It is found almost everywhere in the United States excluding parts of the Rocky Mountains.

The greater wax moth is known as bee moth, wax worm, wax miller, bee miller, or simply wax moth (Figure 3-5). There are

FIGURE 3-5

Wax worms (here in larva stage) can be purchased from biological supply companies. They require little daily care and demand a minimum of classroom space while revealing the stages of an insect's life cycle, throughout the year.

Photo courtesy: Carolina Biological Supply Company

many advantages to raising this insect in the classroom. Wax moths are hardy, odorless insects that do not require daily care; they clearly demonstrate all stages in the life cycle of an insect: egg, larva, pupa, and adult; since it has specialized in bee hives there is little danger of an infestation in the classroom. In addition, extra larvae can be fed to frogs, fish, birds, or other larger animals.

The greater wax moth can be obtained through local bait shops under the name of wax worm; it can also be purchased from Carolina Biological Supply Company or Nasco Company. (Refer to listings at the end of the chapter).

When ordered from a biological supply company, the greater wax moth comes in a prepared medium that should sustain larval growth for the duration of that population. All you need is a large container with a wire mesh top. Since the moth cultures do best when in temperatures around 90°F., you may wish to incubate them. However, if the container is placed near a window and covered with a darkened top (having a few air holes), the temperature may build up to an acceptable range. A light on the side of the container could also substitute for an incubator.

The larvae can survive in a room kept approximately 75°F. (about normal room temperature), but slightly lower temperatures usually result in somewhat slower development; their normal life cycle is completed between four to six weeks. Keep the container darkened so the developing larvae will remain near or above the surface of the medium during development.

You can prepare a medium to subculture these insects by mixing two cups of table sugar with two cups of water and two and one-quarter cups of glycerine. Mix vigorously and pour into a saucepan. Cover and heat slowly for approximately 15 minutes. Allow the mixture to cool and add .6 ml. of a vitamin (Meads) Dica-Vi-Sol. Pour this mixture into five and three-quarters cups of dry Pablum. Mix thoroughly and pour on top of wax moth eggs. The eggs can be obtained by placing wax paper on the surface of the original medium after the adults have emerged.

Another way to subculture or initiate new cultures of the wax moth is to transfer part of the old medium (with larvae) to the new medium.

Questions that might be pursued with this moth are numerous:

1. What stages does the wax moth pass through in its development?
2. How do these insects react when placed near a light (while in a dark room)?
3. Do these insects show a reaction to red or yellow lights (while in a dark room)?
4. How does the moth larvae spin its pupa case?
5. Do temperature differences affect the development of this insect?
6. Does a newly hatched larva move more rapidly than other larval stages? (Observe hatching.) How could this help with survival?
7. How does the jawless adult emerge from the pupa case?
8. How do bees defend their colony against the wax worm?
9. If a few adult wax moths are released in a room filled with light and dark objects, on what objects would they land (in greatest numbers)? How could this help them survive?

When completing the study of the wax moth it is advisable to destroy the cultures. Submerge the adults in Chlorox for several minutes or add extremely hot water to their environment; make certain that all adults are dead. If released they could endanger bee hives in the immediate vicinity, although chances are not likely during most of the school year. Just the same, you might be able to maintain cultures year-round by regulating their numbers in new subcultures. As mentioned previously, these insects will live about a month and a half with no daily care.

For further reference see Dutky, Thompson, and Canwell (1962); NASCO's Teacher's Guide, "The Wax Moth: An Ideal Laboratory Animal for All Grade Levels from Primary Through College"; and Vaughan and Walton (1970).

When completing studies with the wax moth you might consider initiating a study with the organism that sustains the wax moth's existence: the honeybee. Just as wax moths can be sustained in the classroom year-round, so can honeybees.

BRINGING THE HONEYBEE INTO THE SCIENCE LABORATORY

The honeybee is an insect long admired for its active work in pollinating flowers and producing honey for mankind. The honeybee maintains a colonial existence; it has divided up the work load. Worker bees guard the entrance to the hive, nurse the queen, clean the hive, make honey, store pollen, and construct a new honeycomb, in addition to making routine journeys to the fields in search of nectar and pollen. Reproduction for the colony is maintained by the queen and the male drones, both of which are larger than the workers.

The honeybee is fascinating to observe and although it costs a great deal in initial expense ($20.00-$65.00), it requires little care once established; in addition, it may exist year-round for years.

Observation bee hives are offered by a number of biological supply houses. The protective glass sides of these hives allow observation within the classroom while the bees are given access to the outdoors through a small channel or tube.

There are many discoveries to be made through observing and experimenting with bees; you might consider the following questions among your studies:

1. Since bees are colorblind, why do beekeepers paint their hives different colors?

The bee keepers stack their hives; with so many colonies so close together, bees may enter the wrong hive and face death. (Each hive has its own distinct odor and intruders are often killed.) Painting each hive a different color helps bees find their proper home, although it should be remembered that contrasting colors to our eyes may be very similar to a colorblind bee. The bees obviously distinguish between shades of grays.

2. Do bees distinguish scents?

Bees can apparently recognize certain odors, and in many cases their sense of smell is well developed. The bees determine odors with their antennae. The male drones have far more sensory organs on their antennae than do the worker bees; these enable the male to locate the female during mating.

3. Can honeybees taste substances?

It is obvious that bees must taste nectar or sugar solutions, but what about other substances? Provide the bees with dishes of sugar, honey and water. Mix traces of salt, lemon or lime juice, or vanilla extract with certain of the sugary solutions and observe the bees' response. Can bees distinguish salty, sour or bitter substances? What substances do the bees prefer? Use a number of sugars such as table sugar, molasses, grape juice, etc., to see if a preference exists.

4. How do bees communicate?

Students can observe the famous tail-wagging dance of the worker bees within the observation hive. If dishes of honey, sugar and water are set out at specific angles and distances from the hive, you may discover the messages intended by specific dances. Refer to von Frisch (1971), Lindauer (1971), or Wenner (1971) for specifics on the tail-wagging dance. In addition to the dance you may observe that touch and scent are a part of recognizing bees from the same hive. Observe the guard bees as they inspect incoming members.

5. How do bees cool the hive?

When the temperature outside the hive increases during summer, bees may begin to fan the hive. If it continues to grow hotter, they may bring droplets of water into the hive (for cooling purposes). On warmer days set out containers of water and compare their visitation to colder days.

6. What stages does a bee pass through in its life cycle?

You should be able to observe egg-laying and adult stages; observing larva and pupa stages may be difficult unless you open a chamber or two.

7. How do the eyes of a bee differ from our own?

The compound eyes of the bee are the most prominent, but the honeybee also has three simple eyes on its upper "forehead." The compound eyes may have 4,000 facets, and are responsible for almost total vision since a bee acts as if it is blind if the compound eyes are covered.

For further references on honeybees see the previously cited authors, Farb (1962) and Ribbards (1964).

When studying social insects, the ant makes an interesting organism to compare to the bee; they are similar in many respects, in addition to being easily studied in the classroom.

A STUDY OF ANTS

There are about 15,000 kinds of ants and they live in just about every type of location, from sidewalks to trees.

While the ant is a powerful creature for its size, it is loaded with defense mechanisms. Ants can protect themselves by stinging or biting their attackers; some ants have a repellent oil that can be sprayed from the body.

In addition, the ant is well aware of its surroundings. You may find it easy to demonstrate that ants can respond to light, gravity, touch, sound, and chemicals. In addition to body hairs and the very prominent antennae, the ant has taste cells in its mouth region. The ant's compound eyes may contain anywhere from 30-600 facets (anything less than 100 facets is generally considered blind).

Ants feed on nectar, sap, fungi, carrion, seeds, plant-lice (honeydew), soft-bodied insects, springtails, or other materials/organisms. They may search over a million square meters for food.

One of the best places to study ants is in the field. Place a piece of bread smeared with jelly on the sidewalk or grass. It shouldn't be too long before a group of ants are hard at work tearing the material apart. (Determine if ants prefer certain types of bread; give them several choices including whole wheat, enriched white, rye, etc.) Use a dissecting microscope or hand lens to observe the feeding and transporting activities. You can also place paper around the bread and remove portions periodically to see if "trails" were established. Place the sections of paper in a nearby vicinity and observe ants passing over it. Are trails established?

Ants can also be ordered from commercial sources. They are often shipped with an observation nest (refer to listings at the end of the chapter). However, an observation site can be constructed without too much difficulty.

One method involves partially filling a quart jar with pebbles and topsoil, and then adding a few handfuls of dirt containing ants from a natural nesting location. When securing the ants, dig deeply in a radius of about six inches around an ant mound typically found in meadows. Attempt to gather some pupae and possibly the queen; don't take too many workers. Be very gentle with these creatures, and cover the hole once secured. (You can also search for ants under the bark of trees, in rotting logs, or under stones; males and females are usually in the colonies in the late summer and early fall when collecting is best.)

Add the ants to the quart jar and place the jar in a pan of shallow water; this prohibits the ants from escaping. Place a small cardboard square over the jar opening; with a small opening cut in this cardboard, the ants have a surface opening to their nest. This also provides a "feeding tray" where bits of seeds, bread, apple slices, containers of honey-water, or dead insects can be placed and easily observed. Vary the diet of the ants and present them with moderate amounts. Be careful not to disturb the nest once the insects have become established, and do not forget to offer them water (provided in a section of wet sponge or in bread).

Another method of preparing an ant-observation nest involves bracing two sheets of glass (9 x 12 inch range) on three sides with wood framing. Do not be concerned with the glass being escape-proof since the entire structure can be placed in the center of a shallow trough of water. Use a section of red-colored cellophane to cover parts of the glass; ants respond to colored light as to darkness. As in the previous method, add pebbles and loose topsoil; secure ants locally. Keep the nest at a regular temperature, 75°F. if possible, and add a long funnel or thistle tube along one corner of the nest; periodically add water so that it can penetrate into the depths of the site.

Other variations include escape-proof (smaller size) glass aquariums, large glass cylinders, or plaster of Paris (see references).

There are a great number of things that can be observed in the ant colony. Students might consider these questions as well as many others:

1. Do ants sleep?
2. How do ants feed?
3. Do ants feed one another?
4. How does an ant clean itself?
5. Do ants clean their nests?
6. How does an ant use its antennae?
7. How do ants respond when the nest is disturbed? (This should be a field observation done only at time of collecting.)
8. What is the significance of winged ants in the colony?

9. Do ants ever fight one another?
10. How do ants respond when placed in cool temperatures? (Place ants under mild refrigeration.)
11. Can ants from several nests live peaceably together? (Mix ants at time of collecting or later add pupae from a different nest.)
12. How do these insects move from one colony to another? (Make another nesting site available.)
13. Do ants store excess food?
14. Will ants follow ant trails made with 0.5 percent formic acid? For how long?

For further reference see Goetsch (1957), Schneirla (1971), Sudd (1967), Turtox Service Leaflet No. 35, "Studying Ants in Observation Nests," and Wilson (1971).

While a colony of ants is interesting to study, praying mantises are a solitary species that feed on other insects.

PRAYING MANTISES—INSECTS OF IMPORTANCE

Praying mantises often frighten people who are unfamiliar with their appearance or behavior, but these insects don't bother man. Instead, they assist us by controlling the numbers of insects that we classify as pests. Praying mantises are valued as voracious feeders on other insects.

Praying mantises can be obtained in the field during late summer and early fall, or their egg cases can be obtained in the fall and winter from biological supply houses.

Make a cage for the mantis by partially filling a quart jar with weeds, twigs and debris. Small air holes should be made in the top of the jar.

Place the egg case in the jar and keep in a warm place away from direct sunlight. When the young emerge from the egg case in several weeks, feed them fruit flies. Fruit flies can be obtained by setting out over-ripe bananas or ordering them from biological supply houses.

As the young develop, offer them small caterpillars, grasshoppers, roaches, or other insects. These can be used directly if starting with captured adults. It is recommended that the adults be released as soon as the studies are completed (if possible).

When observing the praying mantis you might have students consider a number of questions, including the following:

1. How many organisms does a mantis consume in twenty minutes? One hour? One day?
2. How is the adult mantis different from other insects? How is it similar? Consider the following:

   a. mouthparts
   b. body shape
   c. colors
   d. life cycle
   e. head region
   f. food preferences

3. What organisms will a mantis eat? (Try a wide variety.)
4. How does the mantis capture its prey?
5. Does the mantis prefer light to darkness?
6. How does a mantis grow? (Moults should be seen.)

Praying mantises have been used as agents that control pests, but so have other insects. The green lacewing *(Chrysopa carnea)* is an active predator of aphids, thrips, moth eggs, caterpillars, and red mites. As larvae they may also eat one another. Lacewings and their food supply can be purchased from Rincon-Vitova Insectories, Inc., P.O. Box A, Rialto, California 92376. For further reference, see Butler (1971) and Rincon-Vitova information sheets supplied with cultures.

Another insect used in biological control of insect pests is the tiny parasitic wasp, *Trichogramma.* As an adult this wasp lays its eggs in the pupa stage of moths. The wasp eggs hatch within the pupa case and the larvae consume the developing helpless moth. Instead of moths emerging from the pupae cases, small wasps emerge from holes in the cases. Eggs of this insect can be ordered from Rincon-Vitova at the address previously cited.

When it comes to insect pests, one of those that frequents the summer evenings is the mosquito. While you may wish to see this insect further controlled, this is one creature that can be raised year-round in the classroom.

RAISING MOSQUITOES IN THE CLASSROOM

We often regard mosquitoes as the adult stage that may succeed in obtaining a meal of blood from our bodies. After all, their "bite" is unforgettable. Yet the method in which their beak (proboscis) enters the skin is fascinating; how they can use heat and scents to find us is amazing. Why only certain mosquitoes feed on blood is baffling.

There is much to be learned by studying mosquitoes. Search for larvae in small cans or puddles of water. Keep the larvae in a fingerbowl and make sure a few decaying leaves are in their water. Cover the top of the fingerbowl with a nylon stocking that is supported by a nearby ringstand. Observe over a period of weeks. Refer to Stong (1970) for details.

If mosquito larvae cannot be found locally, they can be ordered from biological supply houses. Their egg cases are also available at certain times of the year.

If winter conditions prevail when you wish to study mosquitoes, bring in a section of soil that often holds summer rain showers. Flood with water and keep in warm temperatures. Mosquitoes may soon appear.

Your students may be interested in solving a number of problems concerning mosquitoes. They may begin by trying to answer some of the following questions:

1. How do mosquito larvae breathe?
2. Of what value are the feeding brushes on the larvae?
3. How many times will an insect molt before emerging as an adult?
4. Do all mosquitoes bite man? (Some males and some species do not feed on man.)
5. Will mosquitoes reproduce without blood in their diets if they normally feed on man?
6. What diseases do mosquitoes carry?
7. Describe the process of an adult mosquito puncturing the skin.
8. How long must a puddle be in existence in order for a mosquito to pass from egg to adult?
9. What organisms eat mosquitoes?
10. Should we attempt to destroy mosquitoes? Why or why not?
11. Where do mosquitoes breed in nature?

For further information refer to Bates (1949), Carpenter and LaCasse (1955), Horsefall (1955), and Stong (1970), and Ward's Culture Leaflet #13.

While people often detest mosquitoes, they also usually dislike flies; their ideas may change if they are given the opportunity to know them both.

OBSERVING THE HOUSE AND BLOW FLY

Flies are not a welcome sight at the dinner table, but they do have a place in nature. This includes the housefly's relatives: fruit flies, black flies, crane flies, horse flies, dung flies, tsetse flies, blow flies, and many other species. While they may transmit diseases and seek the foulest smelling garbage cans, these insects have adapted to specific ways of life. They are a part of the environment, so why not allow them to occupy just a small part of your classroom?

There are many advantages to investigating flies: they occupy little space, require inexpensive and available foods, and can be easily maintained in the classroom. In addition, living specimens can be obtained directly in the human environment or ordered year-round from biological supply houses.

To obtain flies in the fall or spring, simply place fish heads or left-over hamburger, pork, or other meats in damp, loosely folded newspapers. Put these materials in a partially covered trash can (or similar container), and observe over a period of time. You may have success by initially setting the meat or fish directly outside the classroom window, allowing flies to visit; in this instance you would want to later transfer the meat to a moist, dark environment where the eggs could hatch.

Once the larvae (of blowflies) emerge from the eggs laid in the rotting meat, remove a few specimens periodically for observation. A smaller covered container, lined with moist layers of newspaper or paper towels, can temporarily hold the larvae to be observed.

Once the larvae enter the pupae stage, remove them from the original container (which by now is located far away from the human nose). Place the pupae in Petri dishes. Cover the containers and tape the lids on; allow for the seeping of air into the dishes.

Once an adult fly is seen in a Petri dish, begin observation sessions; students should be able to witness the adult emergence as the flies rest within the Petri dishes.

The adults can be transferred to a larger container after emerging. You can immobilize them with carbon dioxide as mentioned previously, or by putting the dishes in the refrigerator for a few minutes.

When studying the adults, a few specimens might be placed in a covered container that has been previously smeared with a thin layer of molasses or honey. Once in this container for a few hours, the wings of the flies may become "glued" down; from here the adults can be placed under dissecting microscopes for closer inspection. (Observation may also be conducted while the adults are within the transparent Petri dishes.)

You may wish to consider the following questions when studying flies:

1. When given a choice of sugar-water or meat, will all flies move to one food source?
2. Offer sugar-water and saccharin solutions to adult flies. Can you fool a fly with a saccharin solution?
3. How does a fly drink?
4. How long does it take a fly to complete its life cycle?
5. What is the reaction of a fly when dusted with fine powder?
6. How does a fly land upside down on the top of a container?
7. What animals feed on flies?

For further information, refer to Dethier (1962).

One advantage of having flies in the classroom is that they can be fed to larger animals when studies have been completed, and so it is with mealworms and flour beetles as well.

MEALWORMS AND FLOUR BEETLES: FOOD FOR LARGER ORGANISMS

Mealworms (*Tenebrio* larvae) can be purchased from many pet stores at approximately one penny per larva; they are also available from biological supply houses.

Place the larvae in a container of shredded newspaper, bran flakes (breakfast cereal), half a potato or apple, and a crust of bread; store in mild refrigeration until needed.

If sustained cultures are desired it is best to place about 25-30 adults in a quart jar filled with the previously mentioned ingredients. Occasionally remove bits of food to new containers of similar materials; eggs present should start new subcultures.

Both larvae and adults provide excellent food for snakes, lizards, birds, frogs, and other organisms; larvae are often preferred to adults.

A number of experiments can be conducted with the hardy larvae. Your students can check their response to light, heat, different foods (they have been maintained on chicken and bird feeds), and other conditions. You can even try to teach these creatures to run a maze or test their response to predators when using them as food organisms.

Flour beetles *(Tribolium)* also make excellent food organisms for larger laboratory specimens. These beetles often infest household flour, so by advising students to bring in specimens when found, you may have a ready supply of insects at your fingertips. These beetles may also be ordered from biological supply houses.

Once securing a starter group, place about 25-30 adults in a pint jar filled partially with flour (all-purpose). Stopper with cotton and keep in a warm place (75°-85°F.) away from direct sunlight.

After about one month remove adults to another flour container; flour beetles eat their own eggs, larvae and pupae. The original flour should contain eggs and larvae; this can be subdivided into new containers, thereby creating more cultures.

Flour beetles have been used in population studies; you might take daily counts of these organisms as they develop in a closed container. Students should find that their numbers do not increase after a certain point; they should observe the adults feeding on their own kind. Attempt variations to the experiments by allowing adults to leave, adding more flour continually, etc.

For further information refer to Carolina Biological culture sheets; Fletcher (1971); Polt (1971); Turtox Service Leaflet No. 34, "The Care of Living Insects in the School Laboratory"; and Ward's Culture Leaflet #13, "Laboratory Maintenance of Common Insects."

TEN ARTHROPODS OF INTEREST

There are many insects and their arthropod relatives that have not been discussed in this chapter and will not be mentioned elsewhere in the book. Some further species to consider are:

Silk-worm	These are available from CCM, but require a fresh supply of mulberry or osage-orange leaves.

Walking Sticks	These are available as eggs from CCM; hatch within the classroom.
Horsefly Larvae	These are offered by NASCO; another fly to compare to house and blow flies.
Grass-hoppers	These may be collected in the fall and placed in a container lined with native soil. After the adults die, place the soil in the refrigerator for a month or longer. Occasionally add water and carefully shake the soil. Remove from the refrigerator and add lettuce to the container. Watch development of larvae. You may have difficulty raising grasshoppers, but why not try it?
Ant Lions	Search for their funnel-shaped traps in fine dust, often at the base of large trees, or in fine sand. Scoop out a handful of sand or dust at the base of the funnel, letting the particles slip between the fingers until the ant lion is revealed. Transfer to a small container of the natural terrain; feed small ants periodically. How do they capture ants? How do they make their funnel traps?
Fruit Flies	Although fruit flies have been mentioned as food organisms for young praying mantises, they are classically used in genetic studies. They are offered by the majority of biological supply houses.
Moth Cocoons	Moth cocoons may be found in your local community in late fall, in trees, shrubs, among dense brush, or under logs. They are also available commercially. Provide an environment similar to natural habitat and tie the cocoons to twigs. In warm temperatures, adults may emerge in 2-3 months. If kept outside during winter, they will emerge when natural foods will be available to complete the next life cycle.
Painted Lady Butter-flies	These insects are available from Carolina Biological Supply Company. They are shipped with a culture medium and require little classroom space and daily care. They provide, throughout the school year, a look at all stages of complete metamorphosis.
Crickets	These insects are available from commercial suppliers, but your best bet is to obtain them locally at bait shops or secure them in the habitat. Crickets can be raised in a mixture of soil and leaves; keep moist (but *not* soaked) and add a crust of

bread, rolled oats, lettuce, and an apple or potato slice. Test their reactions to wooden blocks soaked in different perfumes and/or chemicals. (Will a cricket climb on blocks soaked in these chemicals or how many crickets can be found on such blocks in five-minute intervals? [Polt, 1971]) Also investigate the chirp; you should notice that the wings, *not* the legs, are rubbed to produce the sound. Count the number of chirps in 14 seconds and add 40; this should give the temperature in Farenheit. Test this to see if it is accurate.

Ticks — While not insects, ticks provide excellent representatives of external parasites. They can be secured from local dogs or cats, or by dragging strips of clothing through low-lying brush. Feed them on mice or hamsters occasionally, and store them in plastic vials between feedings or experimentation. *Do not* investigate ticks if in the vinicity of Rocky Mountain Spotted Fever Area. Regardless, be extremely careful when working with them.

Local Insects — In closing this chapter, search in your local vicinity in spring and fall for insect species in abundance (aphids on plant stems, leaf miner caterpillars in trees, etc.). A field trip is an excellent way to observe insects, but avoid wasp or hornet nests; an effective way to observe stinging insects is to watch them feeding on flowers.

REFERENCES

Bates, Marston. *The Natural History of Mosquitoes.* New York: Harper and Row, 1949.

Borror, Donald J. and Dwight M. Delong. *An Introduction to the Study of Insects.* New York: Holt, Rinehart and Winston, 1964.

Butler, George D., Jr. "Techniques for Rearing Lacewings." *American Biology Teacher,* vol. 33, no. 7 (October 1971), pp. 421-423.

Carolina Biological Culture Sheets. Carolina Biological Supply Company, Burlington, N.C. (Accompanies orders.)

Carpenter, Stanley J. and Walter J. LaCasse. *Mosquitoes of North America.* Berkeley: University of California Press, 1955.

Dethier, Vincent G. *To Know a Fly.* San Francisco: Holden-Day, 1962.

Dutky, S.R., J.V. Thompson, and G.E. Cantwell. "A Technique for Mass Rearing the Greater Wax Moth," reprinted from Proceedings in the Entomological Society of Washington, vol. 64, no. 1 (March 1962).

Farb, Peter and the Editors of Life. *The Insects. (Life Nature Library.)* New York: Time, Inc., 1962.

Fletcher, Jack. "Mealworms Galore." *American Biology Teacher,* vol. 33, no. 7 (May 1971), pp. 293-294.

Geltsoff, Paul S. *et al. Culture Methods of Invertebrate Animals.* Ithaca, New York: Comstock Publishing Company, 1937.

Goetsch, Wilhelm. *The Ants.* Ann Arbor: The University of Michigan Press, 1957.

Gorvett, Jean. *Life in Ponds.* New York: American Heritage Press, 1970.

Horsefall, William R. *Mosquitoes: Their Bionomics and Relation to Disease.* New York: The Ronald Press Company, 1955.

Lindauer, Martin. *Communication Among Social Bees.* Cambridge: Harvard University Press, 1971.

NASCO Teacher's Guide. *The Wax Moth: An Ideal Laboratory Animal for All Grade Levels from Primary Through College.* NASCO, Fort Atkinson, Wisconsin. (Accompanies orders.)

Needham, James G. and Paul R. *A Guide to the Study of Fresh-Water Biology.* San Francisco: Holden-Day, Inc., 1962.

Pennak, Robert W. *Fresh-Water Invertebrates of the United States.* New York: The Ronald Press Company, 1953.

Polt, James M. "Experiments in Animal Behavior." *American Biology Teacher,* vol. 33, no. 8 (November 1971), pp. 472-480.

Reid, George K. *et al. Pond Life. (A Golden Nature Guide.)* New York: Golden Press, 1967.

Ribbards, C.R. *The Behavior and Social Life of Honeybees.* New York: Dover Publications, 1964.

RINCON-VITOVA Information Sheets on *Trichogramma,* Green Lacewings *(Chrysopa carnea).* Rincon-Vitova Insectaries, Rialto, Calif.

Schneirla, T.C., ed. by H.R. Topoff. *Army Ants: A Study of Social Organization.* San Francisco: W.C. Freeman and Company, 1971.

Simon, Seymour. *Animals in Field and Laboratory.* New York: McGraw-Hill Book Company, 1968.

Stong, D.L. "The Amateur Scientist: How to Study the Life of a Pond and to Cultivate Aquatic Insects." *Scientific American,* vol. 222, no. 3 (March 1970), pp. 131-136.

Sudd, John H. *An Introduction to the Behavior of Ants.* London: Edward Arnold Publishers, 1967.

Turtox Service Leaflet No. 38, "Moth Cocoons." Chicago: CCM: General Biological, Inc.

Turtox Service Leaflet No. 35, "Studying Ants in Observation Nests." Chicago: CCM: General Biological, Inc.

Turtox Service Leaflet No. 34, "The Care of Living Insects in the School Laboratory." Chicago: CCM: General Biological, Inc.

Vaughan, Brenda and Marcia Walton. *Rearing the Greater Wax Moth. Study Science Aid No. 3.* Washington, D.C., October 1970. (Price: 10 cents from Superintendent of Documents, Washington, D.C. 20402.)

von Frisch, Karl. *Bees: Their Vision, Chemical Senses, and Language.* Ithaca, N.Y.: Cornell University Press, 1971.

Ward, Henry B. and George Whipple. *Fresh Water Biology.* New York: John Wiley and Sons, 1959.

Ward's Culture Leaflet No. 13, "Laboratory Maintenance of Common Insects." Ward's Natural Science Establishment, Inc., Rochester, New York.

Wenner, Adrian D. *The Bee Language Controversy: An Experience in Science.* Boulder: Educational Programs Improvement Corp., 1971.

Wilson, Edward O. *The Insect Societies.* Cambridge: Harvard University Press, 1971.

*AUDIO-VISUALS**

Films:

Army Ants: A Study in Social Behavior (EBEC)
Biography of a Bee (Moody Institute)
Butterfly and Moth Field Series (IFB)
Cecropia Moth (IFB)
Flowers and Bees (EBEC)
Insects Are Interesting (IFB)
Insect Metamorphosis (BFA)
Life Cycle of a Mosquito (McGraw-Hill)
Life of a Butterfly (Wards)
Life Story of a Social Insect (EBEC)
Order of Insects (Thorne)
Pond Insects (EBEC)
Secrets of the Ant and Insect World (Walt Disney)
Social Insects: The Honeybee (EBEC)
The Flower and the Hive (IFB)
The Life of a Dragonfly (BFA)

Filmstrips:

About Insects (Wards)
Honeybee Anatomy and Life Cycle (Carolina Biological)
Insect Life Cycles (5 filmstrips) (BICO catalog)
Life Cycle of Insects (EBEC)
Metamorphosis of Moths and Butterflies (SVE)
Order of Insects (8 filmstrips) (BICO catalog)
The World of Ants (5 filmstrips) (BICO catalog)

Film Loops:

Anthill Protection (I, II) (Doubleday)
Ants–Tunnel Building (Doubleday)

---

*Refer to glossary (page 211) for key to abbreviations and addresses. (Most audio-visuals apply to upper elementary through high school levels.)

Film Loops: (cont.)

Aphids (Thorne)
Bees: Pollen and Nectar Dance (Doubleday)
Butterflies (6 loops) (CCM)
Caddisflies (Thorne)
Caterpillar to Moth (Doubleday)
Dragonfly (Thorne)
Insect Life Cycles (5 loops) (Ealing)
Helpful Insects (Doubleday)
Harmful Insects (Doubleday)
Ladybird Beetle (Thorne)
Mating Behavior in the Cockroach (BSCS)
Mayflies (Thorne)
Mosquito (Thorne)
Pipevine Swallowtail (Thorne)
Queen Bee Duel (Doubleday)
Queen Bee Laying Eggs (Doubleday)
Raising a Queen Bee (Doubleday)
Swarm of Bees (Doubleday)
The Arthropods (6 loops) (Holt)
The Insects (6 loops) (Holt)
Water Tiger Beetle (Thorne)

Also available are BSCS Inquiry slides: Metamorphosis in Cecropia Moths; Control of Molting in Insects (Harcourt)

LIVING INSECTS FOR THE CLASSROOM*

	BICO	*Carolina*	*Conn.*	*CCM*	*So. Bio.*	*Stansi*	*Mogul-Ed*	*Nasco*	*Ward*
Dragonfly nymphs (12)	–	5.50	3.50	–	–	–	–	–	–
Cockroaches (large)	–	2.75	3.00	–	–	–	–	–	5.00
Wax moths	–	3.50	–	–	–	–	–	5.00	–
Honeybee and hive	–	78.00	17.50	72.50	19.95	–	–	17.50	21.95
Ants and nest site	7.95	–	8.50	–	–	–	–	7.95	–
Fruit flies	–	3.50	3.75	4.00	–	3.75	4.50	4.00	–
Mosquito larvae	–	2.75	2.50	–	–	–	–	–	–
Mosquito egg rafts	–	4.50	3.50	–	–	–	–	–	–
Housefly larvae	2.50	2.50	–	–	–	–	–	–	–
Blowfly larvae	–	2.50	–	–	–	–	–	–	3.50 (pupa)
Praying mantis egg case	–	–	–	1.90	–	–	4.80	–	–

*Refer to glossary (page 210) for key to abbreviations and addresses.

LIVING INSECTS FOR THE CLASSROOM *(continued)*

	BICO	*Caro-lina*	*Conn.*	*CCM*	*So. Bio.*	*Stansi*	*Mogul-Ed*	*Nasco*	*Ward*
Moth Cocoons (6)	–	5.00	5.00	5.00	–	–	1.75[1] 1.35[2] 2.00[3]	–	–
Painted lady butterfly	–	5.50	–	–	–	–	–	–	–
Mealworms, *Tenebrio,* 100 larvae	3.00	3.50	2.50	3.50	3.20	5.00	2.20	2.50	3.50
Tenebrio adults (25)	–	5.50	–	–	2.70	–	2.85	–	–
Flour beetles	–	4.00	–	–	–	–	–	–	–

[1]*Polyphemus*
[2]*Cynthia*
[3]*Cecropia*

(Prices in effect at the time this book was written; request organization's current price list.)

While Edmund Scientific Company is not a biological supply company, they do offer praying mantis egg cases and a few miscellaneous materials. Address: Edmund Scientific Company, 615 Edscorp Bldg., Barrington, N.J. 08007.

FOUR

Investigating a Rotting Log-Soil Habitat

While a rotting log slowly decomposes into soil, it harbors a host of organisms (Figure 4-1). Most of the creatures are large enough to be easily observed in the habitat; they are easily collected, and they can be sustained in the classroom with little daily care.

SELECTED INVESTIGATIONS ON THE BEHAVIOR AND ADAPTATIONS OF EARTHWORMS

There are about 2,000 kinds of earthworms; some are as long as 12 feet although four inches is "typical."

An earthworm generally lives from three to 18 inches below the ground surface, although it may burrow down approximately eight feet; it usually burrows deepest in winter, staying below the frost line.

The earthworm is a hardy animal. It can live over eight months without food; it can lose approximately 70 percent of its body water and still survive. Its skin secretes mucous materials that ward off bacteria, molds, and certain predators. An earthworm may live as long as six years, although an average lifespan is estimated at around two years.

Earthworms can be secured from a nearby field/woodland or purchased from local bait shops; night crawlers are the larger specimens while red wigglers are much smaller, although generally

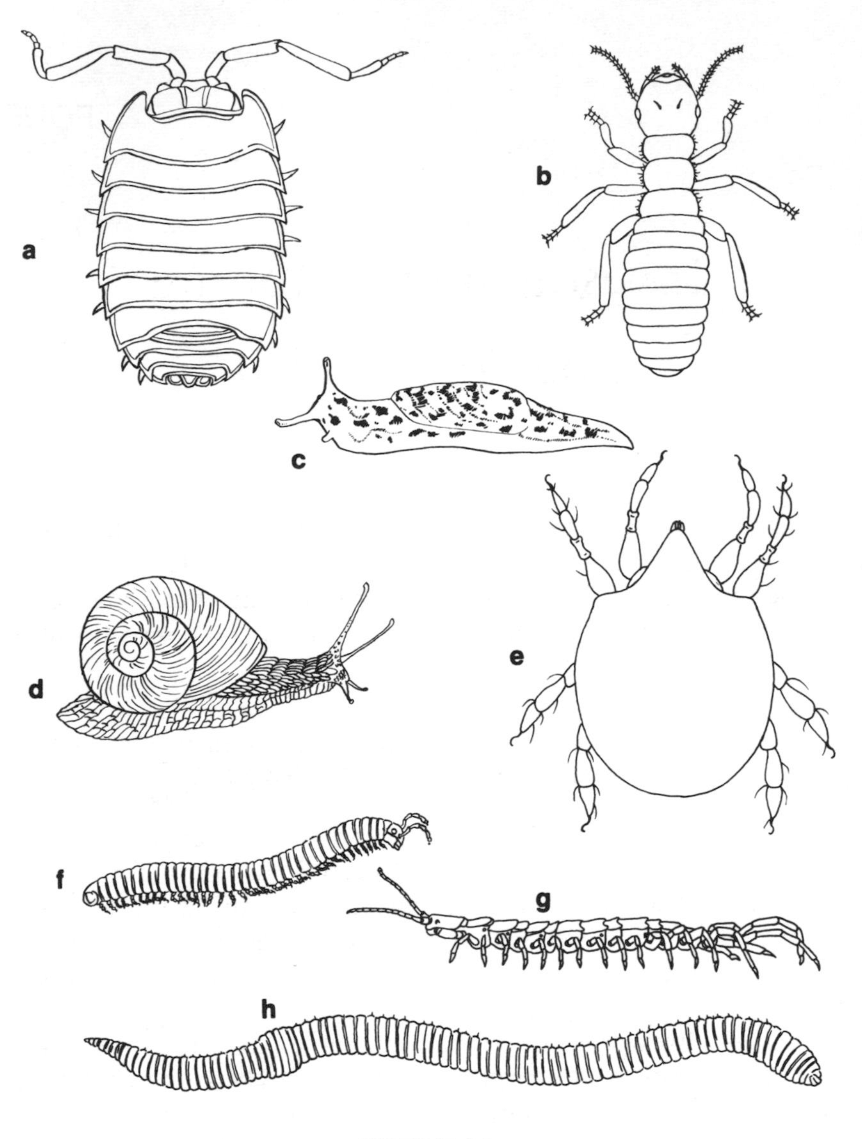

FIGURE 4-1

The following invertebrates can be found in decaying wood, or in the soil: (a) sow bug, pill bug or wood louse *(Oniscus);* (b) termite; (c) garden slug; (d) snail *(Helix);* (e) mite; (f) millipede; (g) centipede; (h) earthworm.

Original drawings by Sheralyn Lerner; after various authors; not to scale.

less expensive per individual. Earthworms are also available from biological supply houses (at more expensive prices than local suppliers).

Earthworms can be sustained in a mixture of natural topsoil rich in bits of decaying leaves and twigs. Keep in a large wooden box and allow part of the area to support weeds or grass. Sprinkle the soil frequently with water, but do not soak the ground. If cultures need to be sustained for long lengths of time, occasionally mix lettuce and manure in the soil.

Earthworms may also do well outside the classroom in a garden compost mound made up of dirt, weeds, grass clippings, leaves, and other debris heaped into a pile. Such a culturing situation is more desirable since the worms thrive in a natural environment. Classroom space is not required and the specimens are always there when you need them; you can borrow from their population anytime you like.

If temporary storage is all that you desire, you may wish to place some worms in a small container mixed with some dirt or sphagnum moss; keep in mild refrigeration.

There are a number of questions or investigations that you can consider with the earthworm:

1. How fast does an earthworm burrow?

Place a worm on top of a container of soil. Use a stopwatch or the second hand of a clock to time the burrowing activity. You can compare burrowing speeds and have "earthworm races." Variables to be considered and discussed are moisture in the soil, size of the worms, temperature, light, etc. Discuss the importance of the burrowing speed.

2. How does an earthworm respond to:
   a. touch of the human hand?

Place an earthworm in a moist, dark container and gently touch at intervals.

   b. attempts to be pulled from the ground?

Allow an earthworm to partially burrow; attempt *gently* to remove it from the ground. What predators duplicate this act in the natural habitat?

c. different chemicals?

Soak diluted household ammonia in a cotton ball; "wring out" the excess fluid and lightly touch to the posterior of an earthworm. Repeat with a weak salt solution, and mild vinegar. Immediately wash the earthworm following contact with the chemicals.

d. light and darkness?

Observe worms placed in a dark container. Cover the end of a flashlight with red, blue, and yellow cellophane wrappings; shine the different colors on the worms. Do worms detect red? (Earthworms cannot "see" red light.)

Determine which part of the earthworm is sensitive to light. Place a pinhole in a piece of cardboard fitted over the end of a flashlight. Shine light on different parts of the worm. Are all parts of a worm sensitive to light? (Anterior portion should be observed as light sensitive.)

e. cold and warm temperatures?

Observe earthworms in mild refrigeration and in varying temperatures slightly above room temperature. Are earthworms more active at higher or lower temperatures?

f. gravity?

Allow earthworms to burrow in a small container of soil. Turn the container at different angles. After a period of time gently search for the earthworms. Do they respond to gravity?

3. Does an earthworm have a heart and veins?

Gently press an earthworm between two glass slides and hold up in front of a bright light (e.g. slide projector). A heart(s) should appear in the anterior portion of the worm with the blood vessels close to the upper and lower surfaces; search for pulsating movements.

4. How does an earthworm breathe?

Have students closely examine the earthworms for structures that might aid them in breathing. Are any structures present? Why must the earthworm's skin be kept moist? (While moisture aids movement and prevents dehydration, it enables air to dissolve into the skin of the animal.) How does the location of the earthworm's veins aid in "breathing"? (They are close to the surface where oxygen can diffuse into the blood.)

5. How does the earthworm move?

Observation of a crawling earthworm should reveal a locomotion involving two sets of muscles; one set runs lengthwise and the other around the body. A hand lens should reveal the tiny "hairs" or setae that help the worm in its movements. These hairs can be felt as rough bristles with the human hand.

For a better understanding of the wavelike muscle movement, cut the last six or seven segments off the posterior end of the worm. Place the two parts together on a section of masking tape. See if a muscle wave will continue to move down the cut portion of the worm if the two parts are touching. Does movement of the earthworm depend on the nerve cord being intact? This is recommended as a demonstration or a larger group activity. Afterwards, use the two parts of the same worm for the next activity.

6. Does an earthworm grow back (regenerate) lost parts?

Place the two previously separated earthworm parts in a Petri dish lined with moist soil or damp sphagnum moss. Count the number of segments on both worm sections. Which part lives? Will segments grow back? (If over 15 segments are removed from the head region of the worm, the back segments will probably starve to death before regenerating a new head region; the earthworm may be able to replace four new segments although it can survive with less than its original number of segments.)

7. How does an earthworm respond to salt?

Use strips of masking tape to initially anchor a large earthworm. Carefully place table salt on the eighth to tenth segments from the posterior end of the worm. How does the worm respond? Continue to add salt to the *same place,* but be careful not to disperse it over the entire worm. What happens? (If the worm's movements spread the salt, quickly wash the worm thoroughly with fresh water and start again with another worm.) This may be best as a teacher demonstration.

8. How do earthworms reproduce?

This is a suitable time to introduce students to the reproductive structure located between the 31st and 37th segments (clitellum). They may also note sexual locations and actual copulation. In natural conditions the earthworms copulate above ground on May evenings. How is it possible for two earthworms to mate and both go away pregnant?

The cocoons formed from the mating act can be ordered from biological supply houses. Place these cocoons in a Petri dish lined with moist filter paper or sphagnum moss. Observe their development over one month.

9. How do earthworms feed?

Students may examine the earthworm's "nose" (prostomium) when seeking an answer to this question. In addition they may examine earthworm casts or observe burrowing and feeding through a glass plate on the side of their container. In its natural habitat the earthworm nibbles everything from dead insects to the tips of pine cones. If a leaf of grass is too tough, it may tenderize it with a type of saliva that will eventually break the substance down. If this doesn't work it may burrow the object so that it will decompose through bacteria or mold action; later it may ingest it when it becomes digestible. The feeding process of the earthworm is a natural introduction to one of the most important values of the earthworm: making topsoil.

10. How does the earthworm enrich the soil?

The casts left behind by the earthworms are loaded with minerals. Soil enriched with earthworm casts has been found to contain about twice as much calcium and magnesium, five times as much nitrates and phosphorous, and over ten times as much potassium, as soil lacking casts. In addition, the burrows and casts enable oxygen and water to enter the soil.

An acre of earthworms has been estimated to produce from 12 to 18 tons of topsoil each year; each worm is estimated to form about one inch of topsoil every decade.

Students may compare soil with and without earthworms, both chemically (with soil-testing kits) and in terms of other life present. Such an experiment should be conducted over a few months at a minimum.

11. Can an earthworm learn?

Construct a maze in the shape of the letter "T." The maze can be built with cardboard, wood, metal, or even hollow glass tubing. On one end of the "T" place a pile of moist, loose soil. On the other end place only bright lights. Allow the worm to "run the maze" several times. Once it discovers the soil, will it turn toward it repeatedly?

Remove the worm and soak the previously used soil with concentrated salt water (table salt and water). Place fresh topsoil in the previously illuminated area. With the light shining on the topsoil, allow the worm to run the maze again. Can the worm adjust to new conditions?

Experiment with other variables: temperature, mild electrical shock, chemicals, different soils, etc.

12. Do insecticides affect earthworms?

This area offers excellent research projects for interested students. Earthworms have been found to store excess amounts of aldrin, dieldrin, heptachlor, and DDT–common insecticides. (Refer to Edwards, 1969.) Could this possibly affect birds that feed on worms? Such insecticide-laden worms could be fed to pond turtles, garter snakes, birds, frogs, or other animals, to trace the effects further up the food chain.

For further reference see the works of Edwards (1969), Farb (1959), Morholt, Brandwein, and Joseph (1966), Savory (1968), Schaller (1968), and Wallwork (1970).

While earthworms make up much of the total weight of soil organisms, the spider relatives often have the greatest number of species.

ADAPTATIONS OF SOIL SPIDERS; ALSO TARANTULAS

The spider-like mites are a predominant type of life in the soil; you can study their diversity (with a hand lens) in soil samples. You may also encounter their relatives–spiders.

Soil spiders are fascinating organisms; they are dark colored and, when still, often resemble bits of debris. Their small size enables them to have a large surface area for their volume; this helps them remain cool when very active. Their body "hairs" and waxy outer coverings help prevent evaporation of body fluids. Their short, thick legs aid in digging activities.

Spiders are predators in the soil community. They have a poison to kill or paralyze their prey. Their jaws or mandibles are designed for sucking out body juices of insects or other victims.

Soil spiders can be gathered from soil samples and maintained in their community; one or two can be further isolated in lesser amounts of *moist* soil. Keep away from direct sunlight and feed small insects periodically.

A soil-dwelling spider of large size is the tarantula that lives in the arid Southwest. While this spider hunts in the evening as do most spiders, it spends the daylight hours in a burrow.

While tarantulas are efficient predators of other arthropods, they do not attack man. Indians have even tied leashes on them and kept them as pets. For its size, the tarantula actually has a mild poison and is not an aggressive creature (in regard to man).

The legs of the tarantula terminate with hooks. The male has special large hooks that are designed to hold the female's fangs during mating. The male is generally smaller than the female and, as with many male spiders, the male tarantula may be eaten by the female during or following mating.

The tarantula is an excellent spider to observe in the classroom because of its enormous size. The general body features are easily observed; the hard sickle-shaped fangs and the cluster of spider eyes on top of the head/chest region are prominent. (Each spider eye focuses on a slightly different region with a different depth of focus.) Tarantulas can be ordered from biological supply houses throughout most of the year.

Tarantulas can be kept in 10- to 20-gallon aquariums covered with fine screening. Partially fill the aquarium with soil, sand or gravel, leaves, and small plants. Feed periodically. Tarantulas may go without food for a week or so if given a large supply of blood, such as from a small or newborn mouse. Tarantulas may also be fed *live* mealworms, crickets, cockroaches, or other insects. A male tarantula may also supply food for the much larger female if placed in the same container.

Study questions can include:

1. What does a tarantula eat?
2. How does a spider "drink" water? (Victim's body fluids.)
3. Does the tarantula spin a web?
4. How does a tarantula breathe?
5. How does a tarantula reproduce?
6. Of what value are the body hairs?
7. How well can the tarantula see with its eyes?
8. How many eyes does a tarantula have?

Other spiders are offered by biological supply houses. Some of these may dwell in the upper leaves and sticks of a brush pile. Compare these spiders to those that dwell in the actual soil.

Make field observations in the sping and fall; record which spiders are present and study their egg cases and web designs. If any spiders are brought into the classroom, release them upon completing your study.

Spiders are not the only hunters in the soil. Another successful predator is the centipede.

OBSERVING MYRIOPODS: CENTIPEDES AND MILLIPEDES

Students may encounter all sorts of many-legged creatures when searching through rotting wood and debris-laden soil. One such animal they may discover is the centipede, often called a "hundred legger."

The centipede has one pair of legs for each body segment. It also has fangs and poison glands to assist in capturing small soil organisms, including slugs.

The centipede's flat shape enables it to live among slivers of decaying wood. The centipede, like most soil creatures, avoids direct sunlight.

Occasionally students may confuse centipedes with millipedes. There are a number of distinct differences although the two types of organisms are closely related.

The millipedes, or "1000 leggers," have two pairs of legs per segment. Incidentally, they never possess 1000 legs; the most legs a millipede can have are 230 or 115 pairs.

The millipedes also differ from centipedes in their way of

life. Millipedes do not have poisonous fangs. They feed on decayed plant life and debris. They depend on special stink glands as their defense; the millipede rolls into a ball when threatened and squirts out a powerful liquid through the glands. This liquid is strong enough to kill small insects, although it will not harm humans.

The millipedes hide under rotten logs, beneath stones, and among the upper leaves of the top soil, although they may go to great depths in the soil. They have two clusters of simple eyes on the front of their heads; these along with the short antennae enable the animals to find their way.

Centipedes and millipedes can be purchased from biological supply houses. While often scarce in winter, these myriopods can be secured in the field for most of the year.

These creatures are easily maintained in a woodland terrarium filled with rotting wood and soil from their natural habitat. (Be sure to return them to their habitat when studies are completed.)

When observing, do not expose them to excessive high temperatures or bright lights. These creatures can be immobilized with carbon dioxide as mentioned with insects in Chapter Three.

Study questions could include:

1. What do these creatures eat?
2. How are centipedes different from millipedes?
3. How do these organisms breathe?
4. How do centipedes and millipedes respond to warmer and colder temperatures?
5. How do these organisms respond to bright light? Can they see all colors that we can?
6. What are these organisms' roles in the soil community?
7. How do their body shapes indicate where they live?
8. How fast can they move in the soil? (Knock out with carbon dioxide and place in a circle of known radius; compute speed.)

An organism that may be found in the soil, living with the centipede and millipede, is the pill bug.

UNIQUE ISOPODS BENEATH OUR FEET

Pill bugs are small soil-dwelling isopods also known as sow bugs or woodlice. These creatures receive the name "pill bug" from their defensive behavior; they roll up into a pill shape when alarmed.

Pill bugs also use their gray to brown color and flattened body shape to remain concealed in their natural habitat. Their home is within or under decaying leaves and wood; they also prefer moist soil. As with other soil creatures, moisture is a necessity. However, the pill bug has special glands on its underside that supply water to the surface in emergencies. This enables the pill bug to breathe while searching for moisture.

Pill bugs can be maintained in the classroom in a terrarium filled with soil and decayed wood from the natural habitat. Occasionally add apple or potato slices to the terrarium. Pill bugs can be purchased from biological supply houses or gathered from the local environment.

While there are a number of features and behavior patterns that could be investigated, you might be interested in the pill bug's response to light. Place electrical tape over one "eye" of the pill bug. Compare his response to a nearby light bulb with one eye taped with his normal response (when both eyes are untaped). Switch the tape to the other "eye" and record the behavior. Pill bugs have been said to have a response to light much like a "tropism," a word often used exclusively with plant responses to light. For other activities: 1. Place soil in a container and completely separate with glass plates sunk into the soil. Soak one portion with water, lightly sprinkle another portion with water, and leave the other one dry. Add pill bugs. Which section do they prefer? Why? 2. Place a light bulb beneath one edge of a metal trough; set the other end within a bucket of ice. Introduce pill bugs. What is the temperature of the location they prefer? Is it the same temperature they prefer in nature? Take the habitat temperature for comparison.

While pill bugs are excellent organisms to study and maintain, snails and slugs are also well suited for classroom study.

A VARIETY OF SIMPLE INVESTIGATIONS CONCERNING SNAILS AND SLUGS

There are about 18,000 kinds of snails in the world; those without shells are called slugs. Snails and slugs expose much of their bodies to the atmosphere. It is understandable why these creatures remain concealed and less active during the day.

While snails are offered by some biological companies, you should be able to find them locally. Snails exist from the deserts to the tropics, from below sea level to mountain peaks. Slugs may be most noticeable in the fall, under moist logs or boards.

Snails and slugs can be maintained in a woodland or garden type of terrarium. Simply add leaves, soil, and a few living plants to an aquarium. Cover with a glass top, leaving a small air space at one corner. Occasionally add fresh lettuce.

There are a number of investigations that can be conducted with snails and slugs:

1. Place a snail or slug on a fresh lettuce leaf. Time its feeding process. How large an area does the animal consume in 30 minutes? Trace the eaten area on a sheet of paper for comparisons; hold "eating contests." Also describe the feeding process. (An aquarium snail feeding on algae on the side of the aquarium may show the teeth action better.) Snails have over 25,000 tiny teeth and they can replace broken teeth. If a snail is placed in a covered shoe box, it may chew its way out, showing the strength of its teeth.
2. Observe the snail or slug in the terrarium. Gently raise one end of the container. Does the worm respond to the change? Eventually, raise the opposite end above the other. Compare results to level ground. Can and does a snail or slug respond to gravity?
3. Allow a slug or snail to crawl onto a glass plate. Observe the underside of the slug and describe the movement. Carefully transfer a specimen to a razor blade or a substitute sharp object. Can a slug or snail crawl over sharp objects without apparent harm? How could this be of value in its natural habitat?
4. Gently tape small objects to the snail's shell. How much weight can a snail carry or pull? (It may pull approximately 200 times its weight and carry 12 times its body weight.)
5. Draw a circle around a snail or slug. Time how long it takes the snail to move from the center to the edge of the circle. Compute the snail's speed. How is this important for the snail's survival? Can it be made to crawl faster?
6. Gently touch a snail's (or slug's) eye; they are located on the tips of long antennae or stalks. How do they respond to touch and how is such a response valuable in protecting the snail's (nearsighted) eyes? Of what value is the location of the eyes on the tips of long stalks?

7. Place a snail in a temporary "dry" habitat (an empty jar with a top having air holes). What is the response? After a few days return to moist conditions and fresh lettuce. How is the snail's response to drought important for survival? (Snails sometimes remain inactive for years when in drought conditions.)
8. Heat several rocks; place several others in the refrigerator and freezer. After a short while place all rocks in a terrarium and take the temperature of the rocks. Add snails and slugs. What is the temperature of the rocks the snails prefer? How do temperatures effect snails and slugs?
9. Search for the slugs' gem-like cluster of eggs that are laid each fall under rotting logs and boards. Do not remove these from the environment; return for periodic observation.
10. Search for snails beneath or within hollow logs or decayed leaves. When discovering several snails, mark their shells with water colors and place them a short distance from their original nesting site. Record their movements. Do snails move about a great deal in the habitat? Do they establish a home?

While overturning rotting logs (and hopefully returning them to their original position), you may notice small plant-like organisms called slime molds.

A DETAILED BREAKDOWN ON HOW TO RECOGNIZE SLIME MOLDS IN THE FIELD: HOW TO RAISE THEM IN THE CLASSROOM

Slime molds are a perplexing group of organisms; they resemble both protozoans and plants.

A slime mold begins its life cycle as a spore that germinates into a microscopic organism something like an amoeba. Eventually two of these protozoans join to form a zygote. The zygote begins to grow and soon resembles a miniature "fan." The fan-like material feeds on bacteria, fungi, and even other slime molds that are in its path. Although it begins microscopically, this fan-like slime (or plasmodium) may grow larger than a human hand, sometimes much larger. Eventually it may give rise to a number of spore-bearing structures. The spores germinate in moist conditions and the life cycle is started once again.

If the fan-like or "slimy" stage (plasmodium) slowly dries out under certain conditions, spores will not be formed. Instead, a resistant stage called a sclerotium (skli-ro-shum) may be formed. The sclerotium may exist for several years during unfavorable conditions and still spring to life when food and moisture return.

During field observations you should search for the spore-bearing stage of slime molds; the spore-bearing stage is the most accessible part of the life cycle; it is also the stage where the specimen can be most positively identified. The spore-bearing stage of slime molds can be found on decaying logs and leaves, although they may exist on tree bark, grass, and other locations. Many species can be found on or in the vicinity of rotting logs.

The spore-bearing structures usually resemble one of these forms:

a. similar to *miniature* "mosses" in appearance; about the size of pins or smaller, having pink, purple, white, brown, black, or yellow tops. Sometimes several tops or spore cases will appear on one stalk. Some of the larger spore-bearing stalks appear as "hair" on the rotting logs.
b. small round spheres, usually red to brown in color. Usually on rotting logs.
c. larger mounds of spores covered with a white to yellow crust. Mature spores appear as a fine dust.
d. whitish twisted masses holding spores on the *outside* of the structures.

Refer to Figure 4-2 for the general characteristics of prominent species; refer to Martin and Alexopoulos (1969) for identification keys.

If you are completely unfamiliar with slime molds it may be wise to invest in a few previously identified samples as offered by the Carolina Biological Supply Company or Southern Biological Supply Company. They offer preserved specimens of *Fuligo, Arcyria, Hemitrichia, Lycogala, Physarum, Trichia,* and *Stemonitis*; these cost from $1.65 to $2.25 each.

If you wish to culture the various slime molds from the spores, be prepared for limited success and close to sterile techniques. Prepare or obtain one of the following types of agar: corn meal, lactose-yeast extract, Turtox lactose, or Knop's solution agar. A "weak" rather than rich agar is necessary; otherwise bacteria may overrun developing slime molds.

Wash the agar with sterilized (boiled distilled) water, and pour off the excess. Hold a recently secured slime mold spore case upside down over the agar. Dust the spores over the agar and

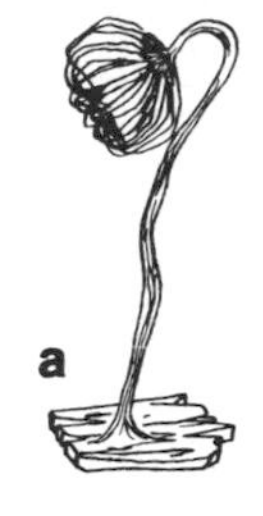

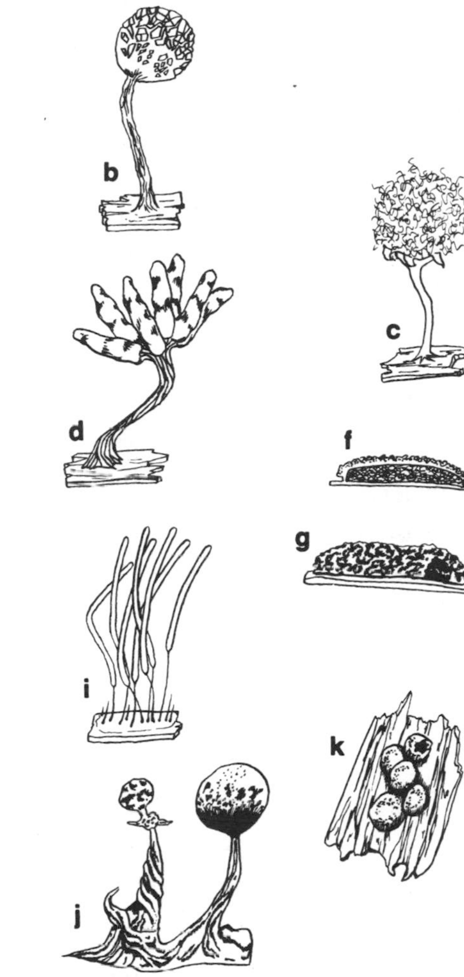

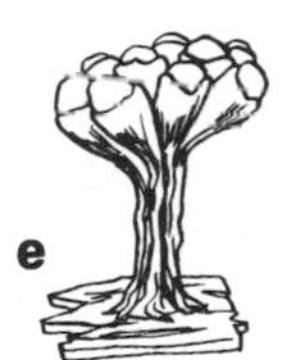

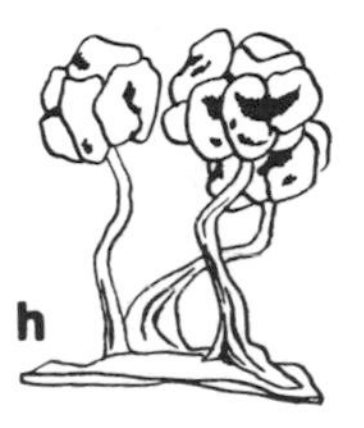

FIGURE 4-2

Many types of slime molds can be found on decaying wood or on decaying leaves; this includes the spore-bearing stages of the following: (a) *Dictydium cancellatum,* **(b)** *Cribraria,* **(c)** *Hemitrichia,* **(d)** *Arcyria cinerea,* **(e)** *Metatrichia vesparium,* **(f)** *Fuligo,* **cross-section, (g)** *Fuligo,* **(h)** *Physarum polycephalum,* **(i)** *Stemonitis,* **(j)** *Didymium* **(often on leaves), (k)** *Lycogala.*

Original drawing by Sheralyn Lerner; after various authors; not to scale.

quickly cover the agar plate. Place the plate in a warm place away from direct sunlight. Allow two or more weeks for plasmodium formation; the plasmodia will be seen as tiny fan-like structures, barely visible at first.

If the spores do not germinate at first and bacteria/mold contamination is not significant, wash the plate again with sterilized water. Sometimes several washings may be necessary, although some spore plates will not germinate in any case. One successful plate in ten is considered about average, although 30 percent success is not unusual.

Once the plasmodium has increased in size, remove it with a section of the agar and place in a moisture chamber. A moisture chamber can be made with a Petri dish lined with filter paper; allow sterilized water to soak into the filter paper but drain away excess amounts of water. After adding the plasmodium, place several rolled oats on the margins of the agar. Rolled oats can be purchased cheaply at a grocery store as old fashioned oatmeal. It would be best if the oatmeal flakes were sterilized in an autoclave or pressure cooker, but heating a test tube of flakes in boiling water may suffice. The slime mold can be taken through its life cycle with the simple addition of rolled oats and adequate moisture.

If field collecting after a spring or fall rainstorm, search for the plasmodium or slime stage. In the field, the plasmodium generally appears fan-shaped, with obvious "veins." The typical color is gold to yellow, although shades of white, violet, and red are possible. The plasmodium is usually found on or beneath loose rotting bark of logs, or among nearby decaying leaves. They have been found beneath corn leaves in winter.

Upon discovering a plasmodium, quickly transfer it to a moisture or humidity chamber and add rolled oats. If the cultures are kept away from direct sunlight the plasmodium may be kept from the spore-bearing stage. Slowly dry it out over several days. This can produce a sclerotium or resistant stage that can be kept in mild refrigeration for several months or longer and yet initiate the plasmodium stage on contact with food and water. Attempt to lure the plasmodium on filter paper with the use of oatmeal, and place enough filter paper in the immediate vicinity so that the plasmodium dries out on the paper. Cut out the dried portions of paper/sclerotium and temporarily store in a pill vial or baby food jar. Keep in mild refrigeration.

Plasmodia can be dried out effectively in bread boxes, large cans, Petri dishes, and other containers. A pie plate, as offered with a graham cracker crust in grocery stores, may be purchased with a molded plastic cover.

Place filter paper on the bottom of the *clean* pie plate; add only enough moisture to soak the filter paper; drain off the excess. Add a growing plasmodium to the paper in the center of the dish; include the agar if present. Place the plastic top on the pie plate and loosely fold the aluminum edging around the base of the cover. Place in a warm dark area for several days. (If the paper is not dry after four days remove the plastic top.)

There are a number of activities involving slime molds:

1. Observe the plasmodium beneath a dissecting miscoscope.

   a. How does it move?

      It has an ebbing and flowing movement where materials stream back and forth; the movement functions something like a muscle contraction.

   b. How fast does the plasmodium move?

      Place finely marked graph paper beneath the agar plate and measure the number of squares the plasmodium advances in 30 minutes. Compute the speed.

   c. How does the plasmodium respond to temperature extremes?

      Place one plasmodium in an incubator (70°-85°F.) and another in the refrigerator. Keep both plasmodia covered (in darkness). Compare size and movements to a plasmodium maintained at room temperatures, also kept in darkness.

   d. How does light affect the plasmodium?

      Place one container in direct sunlight, another in the shade, and a third plasmodium in total darkness. Observe over a period of several days or longer. Compare plasmodium sizes and development.

   e. How does a plasmodium respond to lack of moisture?

      Slowly dry out a plasmodium. One of three possibilities may occur: the plasmodium may be destroyed, the spore stage may be formed, or the resistant stage (sclerotium) may develop.

2. You can also observe spore germination. Place spores of a field-gathered slime mold (*Stemonitis, Physarum,* etc.) in a drop of water on a slide. (Gently dust the spores over the water droplet.) Place a ring of Vaseline Petroleum Jelly around the water drop and add a cover slip. Observe after an hour or two and add more water, if necessary. Germination should take place from one to ten hours after contact with the water. Search for moving cells. How are these similar to protozoans? Adding sodium taurocholate, keeping pH slightly acid, and maintaining 22-30°C. temperatures may increase chances of spore germination (refer to Martin and Alexopoulos, 1969).
3. You may also wish to revive sclerotia that have been formed in the laboratory or obtained from biological supply houses.

   Place the dried stage in a humidity chamber lined with moist filter paper and add several flakes of old-fashioned oatmeal; an alternative method is to place the dried portions directly on an agar plate (corn meal, etc.). How much time is required for the sclerotium to activate? How is this stage valuable in the habitat?
4. Observe spore formation. Continue to feed a growing plasmodium such as *Physarum* that can be obtained through biological supply houses. The plasmodium can be fed oatmeal flakes or agar. In about two weeks the plasmodium should "fruit." Look carefully for initial fruiting at the edges of the plasmodium; once started, fruiting may run its course within hours.
5. Examine a cellular slime mold called *Dictyostelium discoideum* (Figure 4-3). You can observe amoebae (myxamoebae) coming together to form slugs that crawl about slowly on the agar plates and eventually form spores on stalks. Cultures may arrive ready to observe, and transfers for additional cultures can be made by inoculating Turtox Lactose agar with *E. coli*; add a "clump" of spores from the original container. Refer to Kotch (1966) for details.
6. In addition, you may wish to make collections of field-gathered specimens. Keying out the specimens will give you an opportunity to learn about differences in structure among slime mold sporangia, etc. Specimens can be stored in small plastic boxes.

For further reference see the works of Alexopoulos (1962), Bonner (1959), Koch (1966), Martin and Alexopoulos (1969), and Sprague and Einhellig (1971). Carolina Biological also supplies directions for culture of slime molds.

While slime molds make excellent organisms for laboratory study due to their limited requirement of space and materials and their ability to be maintained year-round, mushrooms have similar advantages.

FIGURE 4-3

Spore-bearing stage (sorocarp) of a cellular slime mold, *Dictyostelium.*

MUSHROOM FARMS

Mushroom farms are offered by biological supply companies. The "farms" are usually shipped ready for observation. Little or no daily care is required. Stages of the life cycle can be observed as well as growth and development.

Spore prints can be made by gently cutting the cap off mature mushrooms and setting them on a white piece of paper. Cover with a glass or bell jar. Do not disturb for 24-48 hours. Examine the spore print after carefully lifting off the mushroom cap. Spore colors are important features in identifying mushroom species, but what is the purpose of spores?

Fungi collections can be made in the spring and fall. Such collections can help you to be aware of local species of mushrooms.

A more exciting mushroom for classroom observation is the bracket mushroom, *Schizophyllum.*

SCHIZOPHYLLUM—A BRACKET MUSHROOM THAT CAN BE CULTURED IN THE LABORATORY

Schizophyllum is a bracket mushroom that can be found world-wide on dead trees. It is whitish to gray in appearance. Refer to Christiansen (1965) or Shuttleworth and Zim (1967) for identification features.

What makes *Schizophyllum* so exciting is the ability of the spore-forming layers to remain dormant during dry conditions. The gills roll up and protect these layers. When returned to moist conditions the gills unfurl and produce viable spores.

Search for *Schizophyllum* in the field and, once obtaining a "dry" specimen, embed its upper surface in Vaseline Petroleum Jelly placed on the underside of a Petri dish cover. Place the cover over a bottom portion filled partially with water. Allow it to remain over the water for three to six hours.

Remove the top and substitute it quickly over an agar plate (DIFCO potato–dextrose or corn meal agar). Slowly turn the top, allowing spores to fall on the agar surface. Examine under a dissecting microscope to make certain that spores are present; too many spores is undesirable. Replace the original top on the agar and store away from direct sunlight. Observe development of the mushrooms in the agar.

For best results you may wish to remove one or two spores and transfer them to a sterile agar plate. Observe development over two weeks.

A variety of observations can be made with *Schizophyllum.* Spores can be seen germinating on the agar within 24 hours; primary and secondary myceliums can be seen forming, and gill formation can be witnessed after about one week. You may even have success in using the spores formed in the cultured specimens for future cultures.

For further reference you may wish to see Christiansen (1965) or Shuttleworth (1967); however, the best resource is found in the works of William Kotch, *Fungi in the Laboratory* (1968). Kotch's work will give excellent descriptions and details on *Schizophyllum* as well as slime molds and other interesting organisms.

While bracket mushrooms help decompose wood, they are but one of many decomposing or reducing organisms. One of the most active is the termite.

INVESTIGATING MICROORGANISMS IN THE DIGESTIVE TRACT OF A TERMITE

Termites can be located in most decaying logs; they can also be ordered from biological supply houses. Termites can be easily maintained in the classroom; simply place a portion of rotting wood with termites in a quart jar and cap with a top that has a few small air holes. Add a few moist pieces of fresh wood from time to time.

Termites have a symbiotic relationship with microorganisms in their digestive tracts. Without them the termites would starve. If the microorganisms are removed or killed by higher temperatures, the termite will die. The various microorganisms can digest cellulose whereas the termite by itself can only *chew* cellulose. (Termites are not the only organisms that use microorganisms to aid in digestion: wood-eating beetles, cockroaches, woodroaches, pill bugs, and even cattle have flagellates, amoebae, or bacteria to assist digestion.)

The termites' protozoans can be observed by simply crushing (or decapitating) a termite on a glass slide and adding a cover slip over the juices. Observe under high power. Search for moving organisms.

Study questions can include:

1. How many kinds of protozoans are present in a termite's digestive tract?
2. How do the protozoans swim?
3. How does the termite "aid" the microscopic creatures?
4. How long will the organisms live once outside the termites body? At what temperatures will they survive the longest?

For further information, refer to Yamin (1973).

Other studies can include:

a. Offer different types of wood to a group of termites. Which wood is preferred? Make certain all wood has been partially decayed and equally moist.

b. Cut a hole in a cracker box. Add a few termites through the opening and observe. Open the box after a few moments. Where do the termites prefer to situate? What is their response to light?
c. Mark a small piece of wood with an arrow pointing to the vertical (up). Occasionally turn the section of wood over. Will termites attempt to stay rightside up?
d. Place various insecticides on blocks of wood. Introduce termites and observe. Which blocks will the termites avoid and, over a period of time, which will they return to?

Before ending this chapter on soil creatures it must be added that many more species can be collected and studied, both in the field and in the classroom.

You may wish to gather and study springtails, various beetle and fly larvae, or mites; you may pursue soil-water cultures in search of soil protozoans (soil amoeba are available from some biological supply houses). In addition, you may wish to place sunflower seeds cut in half, in mixtures of soil and water. This may culture fungi in the soil. The studies are endless. (Refer to ants and *Allomyces* for two other soil-dwelling organisms, as discussed in Chapters Three and Six.)

You may wish to construct a Berlese Funnel to gather some of the smaller animals. This involves placing a small screen with fresh soil placed on it over an inverted funnel. At the narrow base of the funnel, place a small baby food jar. As the soil dries out, the smaller organisms will migrate down the funnel into the jar.

If you find the small organisms that migrate into the jar dehydrate immediately upon arriving, you may wish to construct a moisture chamber at the baby food jar destination. Simply place moist filter paper at the capture point. If you wish to preserve the organisms place a diluted alcohol or formaldehyde solution in the jar.

For further information refer to Savory (1968), Yamin (1972), or Wallwork (1970).

REFERENCES

Alexopoulos, C.J. *Introductory Mycology.* New York: John Wiley and Sons, 1962.

Bonner, J.T. *The Cellular Slime Molds.* Princeton: Princeton University Press, 1959.

Carolina Biological Culture Sheets on *Physarum* and *Dictyostelium.* Carolina Biological Supply Company, Burlington, N.C.

Christiansen, Clyde M. *Common Fleshy Fungi.* Minneapolis: Burgess Publishing Company, 1965.

Edwards, Clive A. "Soil Pollutants and Soil Animals." *Scientific American,* vol. 220, no. 4 (April 1969), pp. 88-99.

Farb, Peter. *Living Earth.* New York: Harper and Row, 1959.

Kotch, William J. *Fungi in the Laboratory.* Chapel Hill, N.C. University of North Carolina Book Exchange, 1966.

Martin, G.W. and C.J. Alexopoulos. *The Myxomycetes.* Iowa City: University Press, 1969.

Morholt, Evelyn, Paul F. Brandwein, and Alexander Joseph. *A Sourcebook for the Biological Sciences.* New York: Harcourt Brace and World, Inc., 1966.

Savory, Theodore H. "Hidden Lives," *Scientific American,* July 1968, pp. 108-114.

Schaller, Friedrich. *Soil Animals.* Ann Arbor: University of Michigan Press, 1968.

Shuttleworth, Floyd S. and Herbert S. Zim. *Non-Flowering Plants.* New York: Golden Press, 1967.

Sprague, William R. and Frank A. Einhellig. "Technique for Obtaining Slime Molds." *American Biology Teacher,* vol. 33, no. 6 (September 1971), pp. 359-360.

Wallwork, John A. *Ecology of Soil Animals.* London: McGraw-Hill Publishing Co., 1970.

Yamin, Michael. "Mutualism–A Laboratory Exercise on Termite Protozoa and Cellulose Digestion." *Science Teacher,* vol. 40, no. 4 (April 1973), pp. 42-44.

Yamin, Michael. "Subterranean Termites–Collection and Culture." *Science Teacher,* vol. 39, no. 3 (March 1972).

*AUDIO-VISUALS**

Films:

Culturing Slime Mold Plasmodium (Thorne)
Life in a Cubic Foot of Soil (Coronet)
Life on a Dead Tree (BFA)
Segmentation–The Annelid Worms (EBEC)
Story of Soil (Coronet)

Filmstrips:

Earthworm (Wards)
Fungi and Slime Molds (Classification of Plants, 1 of 9 filmstrips) (BICO Catalog)
Life in a Fallen Log Microcommunity (SVE)
Segmented Worms (McGraw-Hill)
Snails and Slugs (EBEC)
The Segmented Worms (BICO Catalog)

---

*Refer to glossary (page 211) for key to abbreviations and for addresses. Majority of audio-visuals apply to upper elementary through high school levels.

Film Loops:

Black Widow Spider (Thorne)
Centipede (Thorne)
Centipedes, Millipedes, and Scorpions (Doubleday)
Courtship in Spiders (Ealing)
Earthworm (Thorne)
How Spiders Capture Prey: Spiders With Webs (Ealing)
How Spiders Capture Prey: Spiders Without Webs (Ealing)
Isopod (Thorne)
Land Snail (Thorne)
Millipede (Thorne)
Mite (Thorne)
Tarantulas (Doubleday)
Tick (Thorne)

ROTTING LOG–SOIL CREATURES

	BICO	*Carolina*	*Conn.*	*CCM*	*So. Bio.*	*Stansi*	*Schettle*	*Stein.*	*Wards*
Earthworms (12)	3.75	3.50	1.75	4.50	3.20	4.50	4.00	2.28	4.00
Earthworm cocoons	3.00	2.00	–	–	–	–	–	–	–
Tarantula	5.50	4.00	–	–	7.95 (4)	5.00 (4)	–	–	–
Spiders (3)	3.50	–	–	–	–	–	–	–	–
Pill bugs (12)	2.00	2.00	–	–	–	–	4.25 (class of 25)	–	–
Land snails (6)	3.50	–	–	7.05	–	–	–	–	–
Slime molds:									
Physarum	–	9.00 (kit)	3.00	3.00	–	3.00	5.00	4.25	4.25
Fuligo	–	–	–	–	–	4.50	–	–	–
Dictyostelium	3.00	6.00	–	6.00	–	6.00	–	4.25	4.25
Termites (25)	–	5.50	5.50	–	–	–	–	–	–
Crickets (6)	–	2.00	2.50	–	–	–	10.00 (125-200)	2.50 (100)	3.50 (12)
Centipedes (3)	–	5.00	–	–	–	–	–	–	–
Millipedes (3)	–	3.00	–	–	–	–	–	–	–
Mushroom farm or garden	–	6.95	6.95	7.00	–	9.95	–	–	–
Mushroom growth kit	–	3.95	3.95	–	–	–	–	–	–

All prices indicate 1972-73 prices although sometimes more individuals or slightly larger specimens may be provided where prices are comparatively higher. All prices are subject to change; refer to latest catalogs for current prices.

FIVE

In and Around Marine Habitats

Approximately two-thirds of our planet's surface is covered with oceans. Yet, how many living marine organisms did you study last year? Of course, marine organisms have a reputation for being difficult to maintain in the classroom, but once involved with them you may find that they are well worth any additional trouble required. (See Figure 5-1.)

If you live near the East or West Coasts you may have few problems securing many of the upcoming specimens. Nevertheless, be most cautious about allowing students to plunder the coast-lines. Remind them to return all overturned rocks to their original positions and keep only a representative or two of each discovered species. If possible return all specimens upon completing your studies.

If you do not live near salt water, you can still order many common species through biological supply houses. In any case, you will need a classroom-situated aquarium for the housing of marine organisms.

ESTABLISHING A SALT-WATER AQUARIUM IN YOUR CLASSROOM

There are two conditions to consider when handling marine creatures: short-term or long-term maintainence. If you desire to keep specimens for only a few days, a glass baking dish, large fingerbowl, or *clean* photographic tray may suffice. A shallow

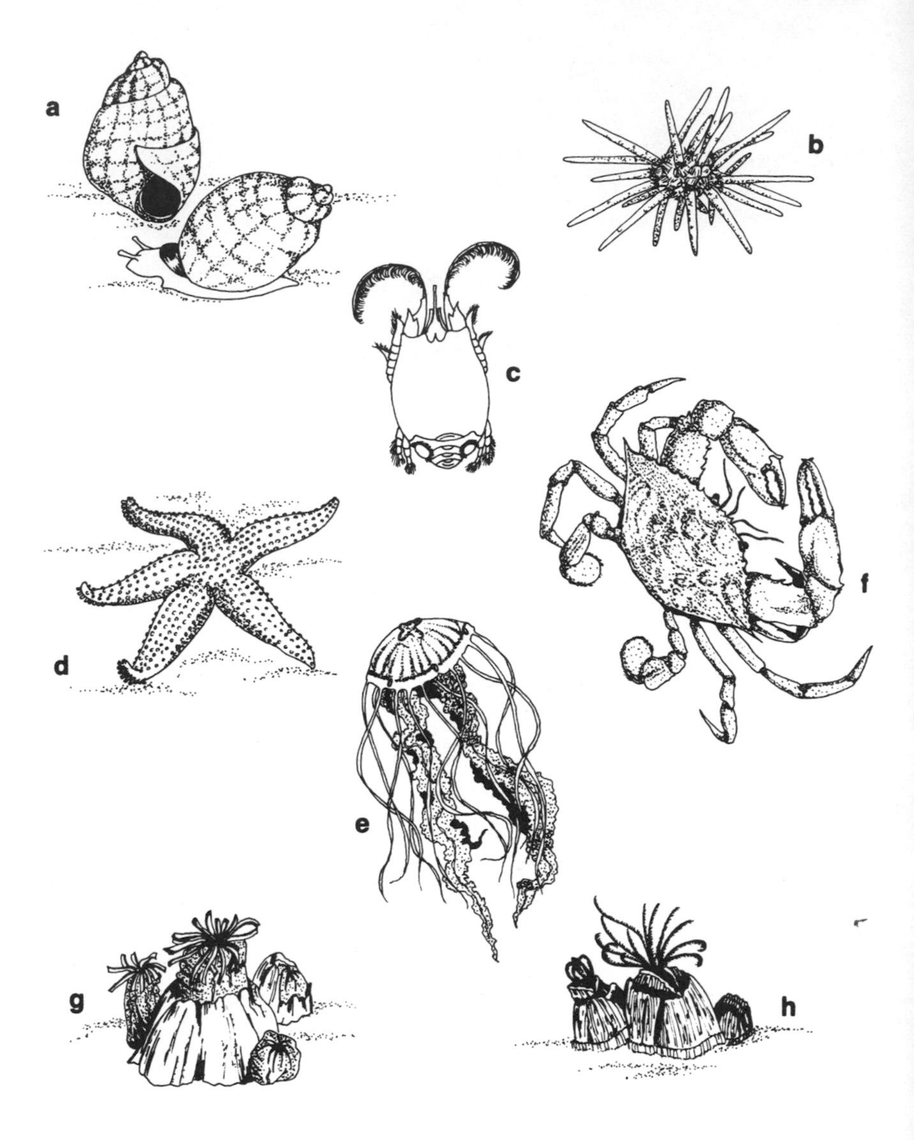

FIGURE 5-1

Marine invertebrates include the: (a) periwinkle *(Littorina)*, (b) sea (pencil) urchin, (c) mole crab *(Emerita)*, (d) sea star, (e) jellyfish, (f) blue crab *(Callinectes)*, (g) sea anemone, and (h) barnacle.

Original drawings by Sheralyn Lerner; after various authors; not to scale.

depth of ocean water or prepared seawater mix will allow enough oxygen to enter the water for a few specimens in each container.

Mark the level of the water with a grease pencil and replace evaporated water with distilled water; keep the container away from direct sunlight and avoid feeding unless part of an investigation.

Some of the marine invertebrates (e.g., crabs) may be sustained for a few days in moist seaweed, under mild refrigeration. If long-term studies are in order you might wish to establish a more permanent home: a salt-water aquarium.

Purchase a 10-20 gallon *all* glass or plastic aquarium; larger aquariums work best in sustaining marine organisms. Avoid metal supports and you can eliminate unsightly corrosion. In addition, you might find some metals form poisonous ions on contact with salt water (e.g., copper).

Line the aquarium bottom with previously washed sand or gravel, perhaps including a *few* shells. Purchase an "under gravel" filter for aeration purposes. (An air stone may work in a two to five gallon aquarium, but an under gravel filter is recommended for all serious marine endeavors.)

Use actual ocean water or a salt-water mix such as can be purchased from biological supply houses or local aquarium shops. Seawater or "ocean water" mixes usually contain sodium, potassium, and calcium chlorides; also magnesium sulfate and sodium bicarbonate. Therefore, table salt dissolved in water will *not* substitute.

While you can mix your own salts (refer to Bayly), unless you plan extensive aquarium work, you should try remaining with actual water from the marine environment, or prepared salt-water mixes.

When collecting ocean water, large heavy-duty plastic bags are usually preferred. Not only do they hold large amounts of water in an uncontaminated environment, but they also store water extremely well.

If an aquarium light is too expensive, several incandescent bulbs may substitute. Turn the lights on for at least 10-12 hours each day.

A thermometer and a hydrometer are important in monitoring temperature and specific gravity of the aquarium; the hydrometer should read between 1.020 and 1.025 at the surface.

If you are working with brackish water, the hydrometer reading may vary somewhat, but record the original reading and compare future readings to this "standard."

If possible, allow the aquarium water to stand for 24 hours before introducing specimens. It might be best to replace a few gallons of aquarium water with fresh seawater at least once monthly. Also, keep a supply of aquarium cement on hand in case of leaks.

For further specifics on maintaining a salt-water aquarium, refer to Bayly or Straughan (1970).

Initially, you should add only one or two *small* fish; feed them tropical fish food or marine flakes. Since overloading an aquarium can be fatal to the inhabitants, be careful as you add a scavenger (small clawless crab or marine snail). A ten-gallon aquarium will probably support about one pound of animal life. While you may wish to add some red or green algae to the aquarium, be sure to remove excessive algae buildup on the sides of the glass.

As with the temporary container, mark the original water level with a grease pencil and replace evaporated water with distilled water, or tap water set out overnight. If you can get a balanced aquarium you may only have to replace evaporated water.

While many marine organisms require a salt-water aquarium while in the classroom, some creatures like the periwinkle can do well in a shoreline environment.

INVESTIGATING THE PERIWINKLE (LITTORINA)

The periwinkle is a gray-shelled snail that can be found clinging to marsh grasses along the coastline. It moves up and down the grass stems as the tidal levels fluctuate. If you cannot gather or observe periwinkles locally, you can order them from biological supply houses.

Periwinkles can be temporarily maintained in a covered aquarium containing a number of moist blades of marsh grass. Since periwinkles are hardy animals that are not extremely shy, they make excellent specimens for study.

You might consider a number of questions involving the periwinkle, including the following:

1. How does a periwinkle respond if:
   a. turned upside down? (Excellent demonstration; repeat under varying conditions and time sequences; compare results.)
   b. placed in a dry container?
   c. placed in mild refrigeration?
   d. placed in a warm incubator?
   e. given a choice of light and darkness?
2. Can you teach a periwinkle to run a "T" maze? Make a "T" shape out of marsh grass and suspend over the table. Add salt at one corner of the "T." Allow the periwinkle to run the maze a number of times.
3. Compare the location of the periwinkle's eyes to those of the garden slug or snail. Why doesn't the periwinkle's habitat (on grass stems) "demand" that the eyes be on the end of long stalks as with the garden slug and snail?
4. Record periwinkle migrations up and down the grasses. Compare their journeys with tidal fluctuations, time of day, day of month, etc. Painting a number or letter on the shell may be necessary for identification purposes.
5. After painting a number or letter on a periwinkle's shell, place it several feet away from its original blade of grass. Will it return to its original location? Trace its movements over several days or longer.

If unable to return periwinkles to their natural habitat, you may wish to crush their shells and introduce them to marine fish or crabs. Periwinkles make excellent natural food for marine specimens.

Living very close to the periwinkles in the marsh environment is a small crab that has a number of interesting adaptations: the fiddler crab.

HOW A FIDDLER CRAB CAN VOLUNTARILY DROP A CLAW

Fiddler crabs *(Uca)* are usually found living on the mud or tidal flats. Their burrows are often numerous, with both males and females on the mud during low tide. You can identify the male by the presence of its large claw (Figure 5-2); the female has two smaller claws.

Fiddler crabs are best studied in the natural habitat, although they can be taken into the classroom for short periods of time. If you cannot secure fiddler crabs from the natural environment, you can order them from biological supply houses.

FIGURE 5-2

A male fiddle crab is identified by its large claw. If a fiddler crab loses a claw, it can regenerate replacements.

Photo courtesy: Turtox

To sustain them in the classroom, line a gallon jar with an inch or so of sandy/muddy material from the natural habitat, if possible. Add enough salt water (seawater mix or from the natural environment) to cover the bottom material to the depth of an *inch.* Leave an elevated area where the fiddlers can always leave the water. Also, add an oyster shell and some red or green algae; keep away from direct sunlight and under room temperatures. Feed small pieces of fish or shrimp to the fiddlers.

There are a number of questions you might consider with the fiddler crabs:

1. What happens if you pick up a fiddler crab by a leg or claw?

You may discover that this crab can voluntarily drop a claw or leg; it has "breaking joints" as do some of its relatives. It can always regenerate replacements.

2. What happens if a male fiddler crab loses its large claw?

You may discover that the smaller claw becomes the "new" large claw, with the replacement claw becoming the smaller of the two.

3. How does the male use its larger fiddle-like claw?

You may discover, in the natural habitat, that the males cannot effectively feed with, or easily carry, the large claw. However, this claw is used to attract females or possibly to defend the territory. In a waving fashion the male fiddler seeks the female's attention. Often a number of males compete for one female; the female enters the burrow of the "winner."

4. If the burrow of a fiddler crab is made of mud and sand, how does it withstand the incoming tide?

The fiddler can seal its burrow and use an air bubble to support the tunnel walls. In some cases the fiddlers make their homes out of a cohesive mud that resists erosion. Regardless, the fiddlers must hide from fish and other predators that enter with the tides.

5. Can a fiddler crab live in a wide range of salinities?

Prepare a number of containers of water, each having a different salinity. Can a fiddler crab live in salinities greater than seawater? How would survival in high salinitites pertain to this crab's survival in shallow tidal pools?

6. Does the fiddler crab change color during the course of a day?

Prepare white, gray, and black paper strips in varying shades. Find a shade that corresponds to the crab's at a specific hour in its habitat. Later (every hour throughout the day), compare the shade to see if the color is the same. How could a color change help a fiddler crab to survive on the mud surface? What could bring about such a change? Does the color change vary with the hour of the day, the tides, the seasons, or the day of the month?

While the fiddler crab's claws can be voluntarily dropped, other crabs can duplicate the regeneration of their claws if lost through a conflict in the environment. In the case of the blue crab's lost claw, it can be closed with electricity, within the confinement of your classroom.

CLOSING A CRAB'S CLAW WITH ELECTRICITY

In this exercise you can investigate the electrical nature of the nervous system of animals. Catch or purchase living crabs (the blue crab, *Callinectes,* works nicely but you might try substitutes).

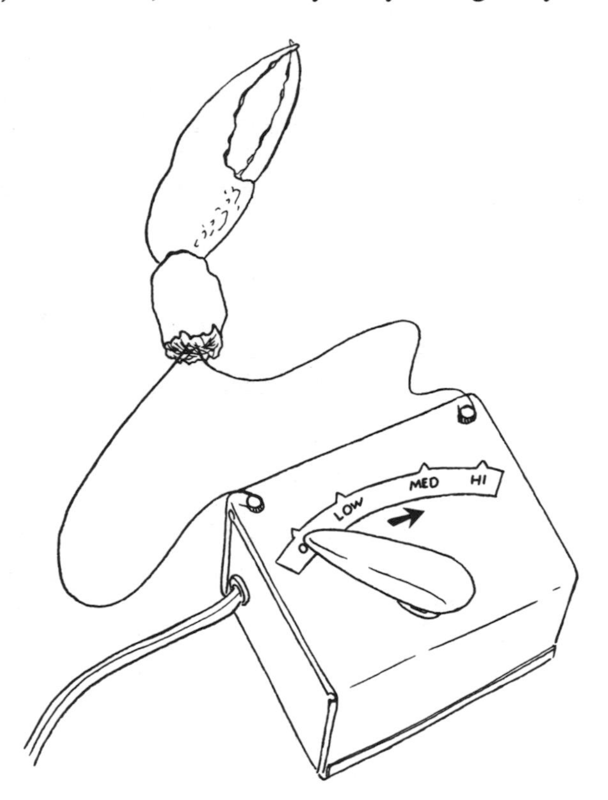

FIGURE 5-3

A toy train transformer can be used to close a freshly broken crab's claw. If the current is slowly raised, more than one response may be observed.

Obtain a simple transformer such as is used with toy train sets. Hook a section of copper wire to each terminal and strip the insulation from the free ends of the wires. With the current turned off, plug in the transformer.

Gently break off a claw at the cleavage plane next to the crab's body. Quickly insert the two copper wires (close together) into the "meat" revealed at the broken end of the claw (Figure 5-3). While holding the wires in place, slowly raise the current with the transformer. At what current strengths are responses noted? How do the responses differ? Is there a relationship between electricity and movement of animal muscles? (For details refer to the College of William and Mary's laboratory Manual.)

Electricity affects animals in various ways; an electrical shock may provide a stimulus for certain sea stars to release their arms.

INVESTIGATING STARFISH AND SEA URCHINS

Since starfish and their echinoderm relatives live exclusively in salt water, you may never discover anything quite like them elsewhere.

Sea stars or starfish come in a variety of shapes and sizes. They may have 25 or five arms; they may have a flexible, serpentine appearance, or be covered with thick inflexible spines. They may be red, purple, orange, blue, or any variety of colors. Nevertheless, they are typically flattened and adapted for dwelling on the ocean's bottom.

You can search for sea stars under rocks or shells along the coastlines. You can also visit fishing docks; fishermen may give you sea stars upon request. In addition, you can always order them from biological supply houses.

Sea stars can be maintained in the classroom in salt-water aquariums or kept for weeks, without feeding, in shallow containers of salt water.

There are a number of investigations that you might pursue concerning starfish:

1. Turn a starfish upside down in a shallow dish of water. How does it right itself? How long does it take? Does it "right itself" faster in bright lights, colder temperatures, etc.?
2. Obtain a "brittle star" (refer to listings at the end of the chapter). Use a variety of batteries from flashlight size to six volt; use these to

give the brittle star varying amounts of electrical shock. What charge is necessary for the brittle star to release its rays or arms? How could such a reaction in nature be of value?

3. Offer a large starfish a section of dead fish or a live clam or oyster. Observe its feeding response. How does its ingestion process differ from other animals? (Be sure to remove all uneaten food afterwards, and change the water.)
4. Use a hand lens to observe the tube feet of the starfish. How do these function? For what purposes are they used? How do they respond to the touch of the human hand? Hot or cold objects?
5. Remove most of one starfish ray. Observe over several weeks or longer. Can the ray or arm be regenerated? Cut a starfish into a number of sections. Which parts grow back into entire animals?
6. Shine bright lights on one half of the starfish's container; provide darkness on the other side. Do starfish prefer light or darkness? How could such a preference relate to their survival?

In regard to sea urchins, these starfish relatives can be found clinging to pilings or rocks along the coastlines. They can also be ordered from biological supply houses.

Sea urchins can be maintained in salt-water aquariums or fairly successfully in trays of salt water. Be careful to avoid the sharp spines when studying these creatures. (Possibilities are slim that an urchin puncture could do much more than cause a mild infection, but using gloves lessens the chance of a puncture.)

A number of investigations involve sea urchins, including the following:

1. Use a hand lens to search for pincher-like structures (pedicellaria) on the urchin; they may also be found on sea stars, although more prominent on some urchins. Sprinkle fine bits of fish food on the surface of the urchin. How do the "pinchers" respond? What is one probable value of these structures?
2. How does the urchin move?
3. How fast does a sea urchin move? Mark a known distance on the bottom or side of the container and record the time required to cover the distance. Convert to feet per second, miles per hour, or in metric units. How is speed important in the habitat?
4. Turn an urchin upside down in a shallow container of salt water. How long does it take the urchin to right itself, and how does it return to the position? Compare the time required when the urchin is placed upside down, next to rocks or large shells.

5. Turn an urchin upside down in a shallow container of salt water and observe the teeth in the "mouth" region. How many teeth are present? How could these be used in feeding among bottom debris?
6. Isolate a number of sea urchins in shallow containers of salt water (50 ml.). Inject two ml. of 0.5 M potassium chloride into the soft membrane surrounding the mouth of the urchins. Observe them over five to ten minutes. Female urchins will give off yellow-orange eggs from their backside (side opposite mouth), while males will give off a whitish mass of sperm on their "backsides."
   a. Compare the sizes and shapes of the two sex cells (microscopically). Which of the two is larger? Which of the two is motile?
   b. Compare amount of eggs to amount of sperm in the containers. Are more sperm or eggs present from an individual? What is the significance of their ratio? In addition, why are so many eggs necessary from each female?
   c. Place a drop of the eggs on a slide and add a drop of sperm. Describe fertilization.
   d. Describe the early stages of a fertilized egg; observe development over 12 hours or longer.

For further details on sea urchin fertilization, refer to Duran (1971) or Harvey (1956). You might also attempt fertilization with sand dollars which can be obtained from biological supply houses. Details of sand dollar fertilization can be found in the work of Wright (1969).

You might also investigate sea cucumbers; these may regurgitate their digestive tracts when alarmed, reveal an interesting feeding sequence, and offer a comparison to other echinoderms studied.

Starfish and their relatives are fairly slow-moving marine specimens. They contrast rapidly moving organisms that can be found living in the wave or surf zone: bean clams and mole crabs.

OBSERVING DONAX VARIABILIS AND EMERITA TALTOIDA: TWO BEACH ORGANISMS IN THE WAVE ZONE

You can use a wide-meshed screen and a shovel to uncover a host of organisms in the wave zone of an ocean beach. Filtering sand through the screen should reveal two prominent specimens: small multi-colored clams and mole crabs. These can also be picked up if you closely observe their digging activity as the waves recede. In addition, bean clams *(Donax variabilis)* can be ordered from biological supply houses.

The bean clams *(Donax)* can be placed in small fingerbowls and studied under dissecting microscopes. You can observe their two siphons and other features.

The bean clam demonstrates best what it must do in order to survive: burrow rapidly. You can fill a tray with moist sand (from the natural environment if possible) and place the clams on the surface. Time their burrowing activity; you can even hold "races" to determine which specimen can burrow the fastest. You might experiment with varying conditions such as bright light, warmer or colder sand, more or less water in the sand, etc. How could these variables affect the burrowing speeds? You might also consider sizes of the burrowing clams or the total number of trials. Relate to the animal's survival in the wave zone where it is continually exposed as waves recede.

Using the same moist container of sand, add more salt water and the mole crabs *(Emerita)*. You can house the crabs in this container as well as conduct "races" here. Again, relate burrowing speeds to survival.

At times in the year you may be able to detect males from females by lifting the flap that covers the "face" of the mole crabs. If an orange cluster of eggs is present beneath the flap you have discovered a female; females are also generally larger. You might study their eggs and attempt to observe the egg development in the laboratory.

Life in the surf zone has adapted to a different type of habitat than many creatures, but all life around salt water has had to adjust to the salt water itself. Salts tend to dehydrate life. Shoreline plants such as the *Spartina* grasses have evolved ways to selectively absorb those salts that will not harm the plant in excess; they have also evolved salt glands to remove excess salts. Tolerance levels fluctuate depending on the organism, from worms to clams.

When it comes to clams, these organisms can provide specimens for experimentation with salinity differences, but they further provide an opportunity to view the beating heart of an invertebrate.

HOW TO DEMONSTRATE THE BEATING HEART OF A CLAM

Obtain clams *(Venus* or *Mercenaria)* from the habitat, fish market, or from biological supply houses. You can keep them alive for a number of days with simple refrigeration. However, when ready to begin experiments, allow the clams to rest for 24-48 hours in a shallow container of salt water similar to the natural environment.

Remove one clam from the salt water and crack the edge of its shell with a hammer; slip a knife through the shell, gently cutting the muscles at either end of the shell. Place the clam in a shallow dish of seawater while *carefully* removing one side of its shell. Locate the heart and allow the animal to "settle down." After a short period of time calculate the number of heart beats per minute under given temperature readings. Place an ice cube in its container and note the change in heart rate. Place different chemicals in the container, such as adrenalin, excessive salts, etc. How do these influence the heart rate? Relate to the clam's response in the habitat.

Prepare a container of sand and salt water; a gallon jar may work well. Add a clam and observe its burrowing activity.

Observe clams in a shallow container of water. You may note that they will open their shells slightly when feeding. What happens if you tap the container when their shells are open? Can light or sound cause a similar response? (You may wish to repeat some of the clam investigations with oysters if clams are unavailable.)

When through with the heart experiments you may wish to offer the clams as food for other marine organisms. While marine fish and crabs prefer bits of fresh clams, you might try feeding them to mud snails.

THE RESPONSE OF MUD SNAILS TO AN OPENED CLAM

You can obtain mud snails *(Nassarius)* along the coastlines or from biological supply houses. Keep in a shallow container of sand and seawater.

The mud snail (or pygmy dog whelk) shows a remarkable

response to food. If you place an opened clam in a container of these snails you can see a chemotactic response first hand. The snails usually show an immediate recognition of the presence of food.

Attempt burying the food or hiding it under or behind objects. Does this discourage these snails? Offer them small pieces of oysters, crabs, or fish in addition to clams. Do they have a food preference? How long does it take them to find the various foods?

Perhaps one of the best-known marine specimens is the jellyfish that often can be found swimming over mud snails at high tide. You might consider bringing the jellyfish into your salt-water aquarium for a brief visit.

OBSERVING JELLYFISH IN AN AQUARIUM

Jellyfish are among the most misunderstood creatures in the sea. People often think they can be increased in numbers if torn apart, or that the reddish color in certain species is due to "blood-sucking." Of course, the jellyfish is just one stage in the life cycle of a coelenterate.

You can gather jellyfish from creeks and bays along the coasts; they are in greatest abundance during the warmer months of the year. (Take along a few large plastic buckets to keep them in and some meat tenderizer to sprinkle on any stung areas contracted while handling the jellyfish.) You can also order jellyfish from biological supply houses.

If you place the jellyfish in a salt-water aquarium you can see first-hand how this creature swims. Describe the motion and discuss its particular answer to locomotion in a liquid environment.

You might smear some of the tentacles on a glass slide and observe some of the stinging cells of the jellyfish. If you can find the attached stage of the jellyfish (on rocks and debris in shallow creeks) you might search for stinging cells in this part of this creature's life cycle. What purposes could the stinging cells serve?

Observe the ability of the jellyfish to stretch many times the length of its contracted position. Estimate the difference between the two positions. How is the stretching possible? What value can it serve? Can other forms of sea life extend to great lengths?

Splash an object on the surface of the aquarium. What is the response of the jellyfish?

Give the jellyfish a choice of a light or dark area in the aquarium. Does it show a preference between light and darkness?

Use ice cubes sealed in plastic bags to lower the temperature of the aquarium. Does this affect the jellyfish? Record responses at various temperatures.

Perhaps as a terminal investigation you might wish to destroy the myth that individual parts of a jellyfish can give rise to new jellyfish. Cut a number of sections of jellyfish and place in individual containers of seawater. Observe for several days, if necessary.

If jellyfish are a problem in your particular locality you might wish to pursue a discussion as to their impact on the local beachers during summer, or why efforts to control them are so difficult.

While searching for the attached stage of the jellyfish, you may encounter another coelenterate that lives on rocks, pilings, and among other debris: the sea anemone.

THE RESPONSE OF SEA ANEMONES TO TOUCH

The sea anemones can be found in shallow creeks or in other salt-water locations. It is best to bring in a section of the material that they are growing on. Transport them in a bucket of shallow seawater; maintain them in a shallow tray of seawater. You can also order sea anemones from biological supply houses.

Sea anemones are especially suitable for two main investigations. First, you can use a probe to gently touch their tentacles after they have relaxed and are fully extended. You should record an immediate response. How is this response of value in the natural habitat? How fast is the response?

A second observation can include the filter feeding process. This is easily seen if the sea anemones are placed in a small aquarium and covered with several inches of salt water; observe with a hand lens through the aquarium glass.

While some of the larger marine invertebrates provide interesting observations, some of the most fascinating organisms, although much smaller, can be found drifting about in the tidal currents: plankton.

HOW TO CONDUCT A PLANKTON TOW

There are a number of ways to conduct a plankton tow, depending on the equipment or facilities at your disposal. It is recommended that you purchase a quality plankton tow net and associated apparatus, although you can make one if necessary. Bend a coat hanger (section) around the wide end of a nylon stocking; cut a hole in the "toe" of the stocking and tie on a small pill bottle. Attach a series of long, but strong, strings to the coat hanger section. (See Figure 5-4.)

FIGURE 5-4

A nylon stocking, a pill bottle, and a bent coathanger with attached strings can make a simple tow net for securing plankton samples.

With your makeshift or purchased tow net, search for a bridge along the coastline, preferably one that has a walkway for pedestrians. Lower the net and bottle into the currents during the late afternoon or evening. Tie the cords to the bridge for about an hour, or slowly walk the net across the length of the bridge, keeping the net at the surface of the water. Later remove the net and examine the bottle contents under a dissecting microscope.

Another alternative is to physically tow the net through the water, using a boat. Either method will work, although the boat will allow you to take plankton tows at various points on a river. Regardless of how you go about it, if you collect a good sample you will have a foundation for grasping the idea of the food chain in marine habitats. A variety of crustaceans and algae may comprise your sample, depending on the gauge of the net and whether or not you used a bottle; the net itself may be used.

While the plankton is comprised of many species, there are numerous marine species that have been neglected in this chapter. Fish will be discussed in Chapter Seven, but hermit crabs, various shrimp, tulip shells, limpets, scallops, bryozoans, and other organisms have been omitted. Use your local abundant species and check the references for additional ideas.

It is hoped that you will avoid octopi and some of the larger, rarer, or more difficult to sustain specimens. It is also hoped that you will further exercise caution when exploring in the marine habitats; overturned rocks should be placed back in their original positions. Even small groups should *not* be taken to the same habitat within short periods of time. As always, you should consider returning species when studies are completed, or observing in the field without taking specimens back to the classroom whenever possible.

REFERENCES

A Reader's Guide to Oceanography. Woods Hole, Mass., Woods Hole Oceanographic Institute. (Free upon request.)

Bayly, Isabel. "Try a Salt-Water Aquarium." Burlington, N.C.: Carolina Biological Supply Company.

Cook, Joseph J. and William L. Wisner. *The Phantom World of the Octopus and Squid.* New York: Dodd, Mead and Company, 1965.

Diehl, Fred A., James B. Feeley, and Daniel G. Gibson. *Experimenting Using Marine Animals.* Eastlake, Ohio: Aquarium Systems, Inc., 1971.

Duran, John C. "Sea Urchin Development: An Investigative Exercise." *American Biology Teacher,* vol. 33, no. 9 (December 1971), pp. 539-542.

Friedrick, Hermann. *Marine Biology.* Seattle: University of Washington Press, 1969.

Harvey, E.B. *The American Arbacia and Other Sea Urchins.* Princeton, N.J.: Princeton University Press, 1956.

Kingsbury, John M. *The Rocky Shore.* Old Greenwich, Conn.: Chatham Press, 1970.

Niemeyer, Virginia B. and Dorothy A. Martin. *A Guide to the Identification of Marine Plants and Invertebrate Animals of Tidewater Virginia. (Educational Pamphlet #13.)* Gloucester, Virginia: Virginia Institute of Marine Science, 1967.

Rabinowitz, Alan, Toby B. Sutton, and Edward M. Taylor. *Oceanography: An Environmental Approach to Marine Science.* Lodi, N.J.: Edward M. Taylor, 1970.

Reid, George K. *Ecology of the Intertidal Zones. (Patterns of Life Series.)* Chicago: Rand McNally and Company, 1968.

Rose, Dale E. and Jacques S. Zaneveld. *Marine Science Field Trip: Educational Series No. 4.* Norfolk, Va.: Old Dominion University (Institute of Oceanography), 1968.

Rudloe, Jack. *The Erotic Ocean.* New York: World Publishing Company, 1971.

Steele, J.H., editor. *Marine Food Chains.* Berkeley: University of California Press, 1970.

Straughan, Robert P.L. *The Salt Water Aquarium in the Home.* New York: A.S. Barnes and Company, 1970.

Teal, John and Mildred. *Life and Death of the Salt Marsh.* Boston: Little, Brown, and Company, 1969.

Wright, R.W. "Fertilization and Embryology of Sand Dollars." *American Biology Teacher,* vol. 31, no. 3 (March 1969), pp. 184-185.

Zim, Herbert S. and Lester Ingle. *Seashores.* Racine, Wisc.: Golden Press, 1955.

*AUDIO-VISUALS**

Films:

Adaptations to Ocean Environments (BFA)
Beach and Sea Animals (EBEC)
Beach and Tidepool Life (IFB)
Between the Tides (Carousel)
Echinoderms–Sea Stars and Their Relatives (EBEC)
Fish in a Changing Environment (EBEC)
Food Chains in the Ocean (BFA)
Life in the Ocean (BFA and McGraw-Hill)
Marine Animals and Their Foods (Coronet)
Marine Animals of the Open Coast (Martin Moyer)
Marine Ecology (McGraw-Hill)
Marine Life (EBEC)
Mysteries of the Deep (Walt Disney)
Survival in the Sea (McGraw-Hill)
The Colorful Cuttle (IFB)
The Restless Sea (Bell)
The Sea (EBEC)

Filmstrips:

Life in the Ocean (PSP)
Life in the Sea Floor and Shore (SVE)
Sponges and Coelenterates (McGraw-Hill)
Sponges, Coelenterates, Ctenophores (EBEC)
The Underwater Environment (Group I, II) (BICO Catalog)

Film Loops:

Brittle Star (Thorne)
Coelenterates and Sponges (CCM)

*Refer to glossary (page 211) for key to abbreviations and addresses; audio-visuals largely apply to upper elementary through high school levels.

Film Loops (Cont.)

Collecting Plankton (CCM)
Hermit Crab (Thorne)
Life in the Intertidal Region (BSCS)
Marine Ecology (6 loops) (Ealing)
Octopus (Doubleday and Thorne)
Predation and Protection in the Ocean (BSCS)
Sand Crab (Thorne)
Sand Dollar (Thorne)
Sea Anemone (Thorne)
Sea Cucumber (Thorne)
Sea Urchin (Thorne)
Shore Crab (Thorne)
Starfishes (Thorne)
Sunflower Star (Thorne)
Survival on the Coral Reef (CCM and Doubleday)
The Intertidal Region (BSCS)
The Undersea World of Jacques Cousteau (13 loops) (Ealing)
Tidepool Life (CCM)
Tidepool Life (Parts I, II) (Doubleday)
Trawl Net Collecting (CCM)

LIVING MARINE SPECIMENS

	Carolina	*Connecticut*	*So. Bio.*	*Turtox:CCM*	*Steinhilber* NASCO
Periwinkle (*Littorina*)	–	4.00 (3)	–	3.00 (12)	4.02 (3)
Fiddler Crab (*Uca*)	–	–	8.50 (3)	7.50 (12)	–
Blue Crab (*Callinectes*)	–	12.00 (3)	–	3.50 ea.	12.00 (3)
Sea stars (starfish)	6.50 (5)	10.00 (3)	–	1.00 ea.	10.02 (3)
Brittle stars (starfish)	–	–	8.50 (3)	7.50 (12)	–
Sea urchins	6.50 (5)	14.00 (3)	–	25.00 (24)	–
Sand dollars	–	–	–	1.00 ea. 10.00 (12)	–
Sea cucumbers	8.00 (5)	–	–	1.00 ea.	–
Bean clams (*Donax*)	–	–	–	7.50 (12)	–
Hard clams (Venus)	–	4.00 (3)	–	–	4.02 (3)
Mud snails (Nassarius)	–	–	–	5.00 (12)	–
Jellyfish	–	–	–	5.00 ea.	–
Barnacles	8.00 (5)	4.00 (3)	–	5.00 (12)	4.02 (3)
Hermit crabs	5.25 (5)	4.50 (3)	8.50 (3)	2.50 ea.	4.50 (3)
Grass Shrimp (Palaemonetes)	–	–	–	5.00 (12)	–
Tulip shell (Fasciolaria)	12.00 (5)	–	–	3.00 ea.	–
Limpets	–	4.50 (3)	–	3.00 ea.	4.50 (3)
Scallops	–	–	–	3.00 ea.	–
Bryozoans	–	–	–	5.00 ea.	–
Sand worms (Nereis) (or clam worms)	–	12.00 (12)	–	10.00 (12)	–
Salt-water mixes	5.80 (8½ lb)	–	–	5.15 (5 lb.)	–

When ordering these marine specimens you will most likely encounter a minimum order requirement (e.g. $25). Prices were in effect at the time of this writing; write companies for latest offerings and prices.

SIX

Selected Plant or Plant-like Organism

Plants or plant-like organisms have a great deal to offer. Their growth rates may be phenomenal; their sizes, microscopic to enormous; and their characteristics, varied and fascinating. In addition, many species can be easily maintained in the classroom.

While it is not intended that you consider plants separately from animals, it is hoped that you might include some of the following plants or plant-like organisms in your studies:

BRINGING BACTERIA INTO THE CLASSROOM

There are a number of reasons you might wish to study bacteria. They require little classroom space; they represent one of the smallest and most numerous types of life on earth; they are easily obtainable; they can be studied for growth rates and decomposing values, as well as a host of cellular activities.

However, there are disadvantages. Bacteria usually require special and certain expensive equipment; Umbreit (1964) estimates a *minimum* of $300 a year to seriously study bacteria. The two essential items are a pressure cooker or autoclave and an oil immersion microscope. Even if the equipment is secured, teacher training is recommended; so we shall remain with simple or more easily implemented ideas. (Sterilization and transfer procedures will follow.)

1. Blend a potato with water and pour into a beaker. Set in a warm, dark place. Examine after a day or two. Are bacteria present? Describe them. You might attempt to take general counts of the bacteria over a number of days. Do their numbers continually climb? What could cause the numbers to decline? This investigation can be repeated with apples, pears, grapes, and other fruits. In addition, the blended materials can be sterilized and used as culture media.
2. Mix a small amount of soil and water in a Petri dish. Cut a few sunflower seeds in half and add them to the container. Cover the container and inspect for bacteria (and fungi) in a few days. What types of bacteria are present: spherical, rod, or spiral shapes?
3. Bring in a small sample of polluted water and examine for bacteria. Are certain bacteria commonly found in polluted water? Where would some of these bacteria normally live?
4. Purchase or prepare Petri dishes filled with agar. Expose different plates to various parts of the school environment: cafeteria, bathrooms, library, clinic, etc. Expose each plate for ten seconds. Attempt to identify which places contain the greatest amounts and numbers of bacteria.
5. Secure a major representative of each type of bacterium; three suitable bacteria that do not cause diseases (non-pathogenic) are *Bacillus subtilis* (rods–see Figure 6-1), *Sarcina lutea* (spheres), and *Rhodospirillum rubrum* (spirals–see Figure 6-2). These can be seen by using microscopes with the highest magnification of light. Refer to upcoming transfer techniques concerning handling of such bacteria.
6. Purchase an antibiotic sensitivity kit, or antibiotic discs containing Penicillin, Streptomycin, Neomycin, Tetracycline, Novobiocin, Erythromycin, and Chloromycetin. Inoculate an agar plate with *Bacillus cereus,* or other non-pathogenic bacteria. Carefully place the various discs on the agar surface and cover quickly. Observe over several days, keeping the plates in a warm dark place during the interim period. Are there clear areas around some of the discs? What does this mean? Are some antibiotics more effective than others against certain types of bacteria? Relate to the use of antibiotics in medicines.
7. Purchase a luminescent bacterium, *Photobacterium fischeri,* and transfer to accompanying plates or slants (slanted agar surface in a test tube). Observe in darkness. What could cause a bacterium to glow in the dark? From where does it get its energy?
8. Purchase or prepare a starch agar medium. Expose to the air or inoculate several plates with a known bacterium. After the agar has been covered with bacterial growth taking up about 50 percent of

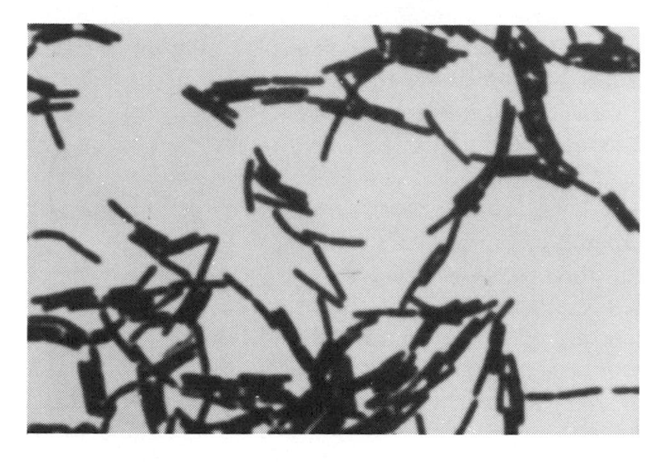

FIGURE 6-1

(Bacillus subtilis) **These bacteria can be purchased through most biological supply houses, and are suitable for classroom studies.**

Photo courtesy: Connecticut Valley Biological Supply Company

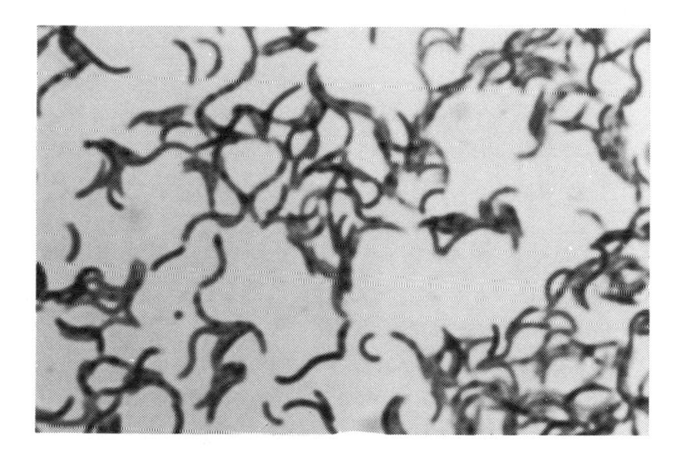

FIGURE 6-2

(Rhodospirillum rubrum) **These bacteria can be purchased through most biological supply houses, and are suitable for classroom studies.**

Photo courtesy: Connecticut Valley Biological Supply Company

the agar surface, add potassium iodide-iodine solution (Lugol's solution). Wash over the entire plate. (This solution will stain starch blue-black.) Compare this to the appearance of a sterile starch agar plate washed with the same solution. Do bacteria feed on agar? On what foods do many types of bacteria feed?

9. Bring in freshly picked clover or bean plants with roots intact. Examine the roots for the presence of nodules. Crush a nodule and add a small amount of water. Examine a drop of this water beneath the microscope. How could bacteria help plants to survive? How could bacteria help all living organisms to survive?
10. Expose an agar plate to the air and, at the same time, expose another plate that has been recently washed with a common disinfectant (Lysol). Compare the bacterial growth that follows. Why is a disinfectant of great value in the household? What effects do soaps have on bacterial growth?
11. Inoculate four plates with the same bacterium and store them in a number of conditions. Place one plate in the refrigerator, another in bright light at room temperature, a third in darkness at room temperature, and a fourth in a warm situation. Record all temperatures and conditions. Compare the growth rates. What conditions do bacteria prefer? Are more investigations necessary to answer this question?

Sterilization and Transfer Techniques

To begin, *treat all bacteria as if they are pathogenic or disease causing;* such species could contaminate your plates accidentally. Do not be careless with bacteria; keep your hands away from your mouth while working with bacteria and do not allow immature students to handle cultures directly. Pour Lysol on any broken test tubes or plates; allow it to remain undisturbed for 20 minutes before cleaning up the spillage. It is further advised to always scrub down the immediate area prior to, and following, the handling of bacteria; use Lysol or 70 percent alcohol.

To sterilize media and glassware prior to or following an investigation, pressure cook or autoclave at 15 lbs. per square inch for 15 minutes. (This same procedure can be used to destroy unwanted cultures.) You can also sterilize glassware in the oven at 320°F. for two hours.

It is best to avoid *all* pathogenic bacteria, even if they are plant pathogens. Unless you are an expert in bacteriology, you may find an ounce of prevention in this case may be worth far more than a pound of cure.

When preparing to transfer one culture to another location, purchase an inoculating loop (or needle). You can make an inoculating loop by securing nichrome wire and inserting a six-inch section firmly into the eraser of a pencil. Twist the free end into a tight loop (Fagle, 1961).

Place the loop into the flame of a Bunsen burner and heat the entire length of the wire until each part has become "red" or glowing. While holding the needle in the right hand, quickly remove the caps or stoppers from two test tubes: one that contains the culture; the other, the sterile agar for new growth. Use the right hand to loosen the caps, and keep the tops between the fingers.

Quickly pass the mouths of the two test tubes through the Bunsen burner flame several times. Keep the open ends pointing downward or parallel to the ground so that spores will not settle on the agar.

Insert the loop into the tube containing the bacteria to be transferred. Allow the loop to touch the medium momentarily to lose excess heat; then gently drag the loop over the agar surface, making contact with the bacterial mass. Quickly remove the loop and insert it into the second tube, gently smearing the material in a wavy pattern on the *surface* of the agar.

Remove the loop, cap both tubes, and immediately flame the needle once again to the glowing state. Be sure to label the new culture.

When working with Petri dishes, you will have to carefully open the top to make the streak on the plate. (Be sure to scrub down the nearby area with Lysol if the plate is resting on a table; working under a hood further aids in preventing contamination when working with plates.)

If transferring to a slide, simply touch the loop to the slide, cap the tube or plate, and flame the needle. You may wish to add (to the slide) a minute drop of water prior to or after adding the bacteria. Add a cover glass and observe under high power on the microscope.

You may also wish to stain bacteria. Permanent inks may suffice although a wide variety of stains are offered commercially.

To sustain bacteria cultures you may wish to transfer a portion of the old cultures to previously sterilized media every few weeks. Refrigerate cultures to extend their life. Refer to Ward's

Culture Leaflet No. 23 for further details on transfer or sterilization techniques.

When securing known bacteria, you may wish to inquire at nearby universities. If you provide sterile plates they may make a few transfers from their stocks, or even give you old cultures; in addition, you might ask your local health department for old plates of *identified non-pathogenic* bacteria. These are often disposed of periodically.

Recommended bacteria include: *Bacillus cereus, Bacillus subtilis, Enterobacter aerogenes, Micrococcus luteus, Proteus vulgaris, Pseudomonas fragi* (heat sensitive), *Rhodospirillum rubrum, Spirillium serpens, Spirillium volutans, Streptococcus lactis,* and *Vibrio percolans.*

Bacteria are microscopic organisms usually found in large numbers; they are often saprobes, existing on the products of living or once living organisms. This description also fits the water molds.

A DETAILED INVESTIGATION OF THE WATER MOLD, ALLOMYCES; HOW THIS ORGANISM CAN BE MAINTAINED INDEFINITELY IN THE CLASSROOM

Allomyces (Figure 6-3) is a water mold that has often been neglected in the classroom. Yet this soil- and water-dwelling

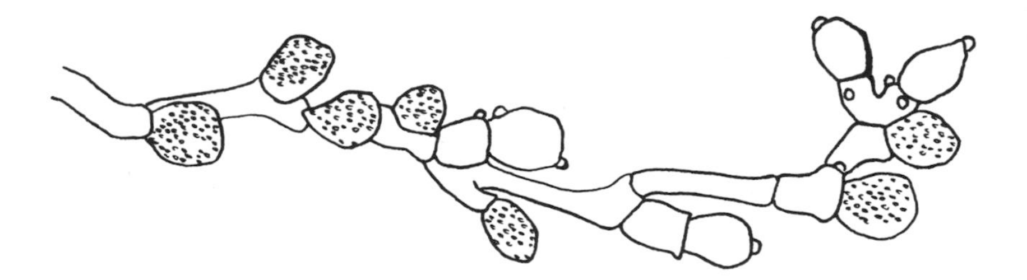

FIGURE 6-3

Sporophyte stage of *Allomyces arbuscula,* a water mold that can be raised on sunflower seeds. (a) Resistant sporangia that can survive environmental hardships.

fungus reproduces with both sexes and spores (zoospores) in its life cycle, and can demonstrate alternation of generations throughout the year. In addition, it provides a representative water mold that not only can demonstrate "swimming" spores, but provides a direct view of fertilization. Perhaps the characteristic of greatest importance is its ability to form resistant spore structures; this can easily enable you to keep this organism in your classroom for years. *Allomyces* can fortunately be maintained on a handful of inexpensive sunflower seeds.

The field method of obtaining *Allomyces* involves gathering approximately 10-15 samples of soil from varying locations. Place a spoonful of dirt in a Petri dish and add distilled water; include a few sunflower seeds, each cut lengthwise in two sections. Place the exposed portion of the seeds face down in the soil-water. Cover and label the collection site and date of preparation.

Observe after several days, continuing through several weeks. If a cottony growth develops around certain seeds, examine beneath the microscope to confirm the identification of *Allomyces*. *Allomyces* has forking branches and cross-walls in its mat of hyphae. You may also note the presence of orange, male structures situated above female sites if you have the sexual stage in the life cycle, or dark, numerous brown spore-bearing structures if you have the sporophyte stage.

Allomyces can also be obtained from biological supply houses; refer to listings at the end of the chapter.

You can maintain *Allomyces* in commercially available Emerson YpSs Agar, Cantino PYG Broth, or on Corn Meal-Dextrose-Peptone Agar (CMDP agar). Kotch (1968) recommends 17.1 gm. of Corn Meal Agar, 8.0 gm. of Dextrose, 1.8 gm. Peptone and 900 ml. of distilled water in the preparation of CMDP agar.

However, the simplest way to maintain *Allomyces* is through the use of sunflower or sterile hemp seeds. Simply cut each seed in two lengthwise sections, using a sterilized single-edge razor blade. Place the exposed side of the seed face down in a shallow (Petri) dish of water. Cover and transfer each circular growth as it expands to four additional seed halves and fresh water in another Petri dish. Use a sterilized wire loop to make the transfer.

As long as transfers are continued regularly there is little worry over acidity becoming a problem, but to help with this problem, keep only four (or fewer) seed halves in each container.

If the mold is allowed to grow for several weeks after a transfer or two, and it becomes increasingly dark in color, it is time to consider removing the seeds with their circular mold growths. Use a sterilized wire loop to transfer the masses to filter paper in a covered container (e.g. two aluminum pie plates with one inverted to form a cover). Allow the water mold to dry out slowly. Once completed, cut away the excess filter paper and store the dried clusters in a covered Petri dish. These may be viable for years.

When ready to activate the dried spore clusters, simply place three sections in a Petri dish; fill partially with water and add several sunflower seeds cut in half as mentioned previously. Transfers can be made later.

The use of non-sterilized sunflower seeds and/or other makeshift adaptations means that your cultures may become contaminated with another fungus or bacteria, possibly both. If this happens you may try washing the contaminants away with sterilized water and beginning new cultures. You may simply wish to abandon certain dishes. In some cases contaminants will leave the scene when the drying stage is undertaken. Nevertheless, for general purposes, a few minor contaminants will not be a problem.

A number of seeds may be used when culturing this water mold. Kotch (1968) recommends cucumber, radish, and pumpkin seeds as alternatives to hemp and sunflower seeds. You may experiment with these and other seeds, although sunflower seeds seem to be the least expensive and are readily available. Offered as bird food, sunflower seeds cost less than a dollar for two pounds at a grocery store. In addition, they are easy to cut lengthwise when compared to hemp or radish seeds. (It should be noted that if the seed is allowed to remain within its shell covering, there is greater chance of success.) Refer to Kotch (1968) for details.

Investigations concerning the water mold, *Allomyces,* can involve the following:

1. Collect the soil samples, as mentioned earlier, and note the location of each sample. Follow the previously stated instructions in using sunflower seeds, soil, and water in Petri dishes. Determine where *Allomyces* exists in your immediate environment. Repeat at various times in the year to determine the best time to conduct yearly studies.

2. Examine *Allomyces* in various stages of growth. Dry out various stages of its development. Which stages, and in particular, which structures in its life cycle, have the ability to survive environmental hardships?
3. Obtain the dried sporophyte stage (brown masses dried out on filter paper). Subject various conditions–strong light, low and high temperatures, salts, physically crushing weights–on the inactive spore structures. How do such conditions affect the mold when returned to seed halves in fresh water?
4. Using the dried sporophyte stage, add a dried mass (thallus) to a Petri dish having both distilled water and a few seed halves. Observe over one month, with frequent observations during the first week. Attempt to answer the following questions:
   a. When do the seeds first show signs of growth of the water mold?
   b. How are the spores (2nd-3rd day) released from the sporangia or spore-bearing bodies?
   c. When do sexes become apparent? (2-8 days.) How do the sex cells differ? Are male sex cells normally smaller in other organisms? How do both cells move?
   d. Describe the fertilization process that follows the development of mature sex cells. Is this typical throughout nature?

When the sexual stage has reached maturity it is time to transfer the molds to new Petri dishes with seed halves. Allow the seeds to become infested with the water mold, observing over three additional weeks' time. Attempt to answer the following questions:

   e. How long does it take for the sporophyte stage to develop spore-bearing bodies? How do these structures differ in appearance from the sexual stage structures?
   f. Are all spore-bearing structures the same in appearance?
   g. Remove the water mold after about three weeks when it is deep brown in coloration. Follow drying procedures mentioned previously. Do all spore structures survive the drying procedure?

Allomyces is but one water mold that you can use in your studies; *Achyla* is another that you may consider. You can place apples (punctured repeatedly with small pins) in cloth sacks, and carry them down to the shorelines of local lakes in an attempt to "bait" various water molds. Leave the cloth sack with apples among the shoreline debris for one to two weeks before retrieving specimens. You may also lower sacks to various depths to determine preferred depths of water molds. Examine in the laboratory.

When working in the lake waters you may discover certain

other plants that can move about freely in the water for more than a part of their life cycles. The plant colonies may stand out beneath your microscope.

COMPARING VOLVOX TO OTHER MICROSCOPIC (PLANT) COLONIES

Volvox is a large hollow sphere of interconnected cells that function as a colony (Figure 6-4). Certain cells guide the colony toward light and certain cells reproduce for the group. Because of its large size and fascinating shape and movement, *Volvox* is well known in the biological world. Yet, to grasp the significance of its colonial existence, you might wish to examine other plant colonies.

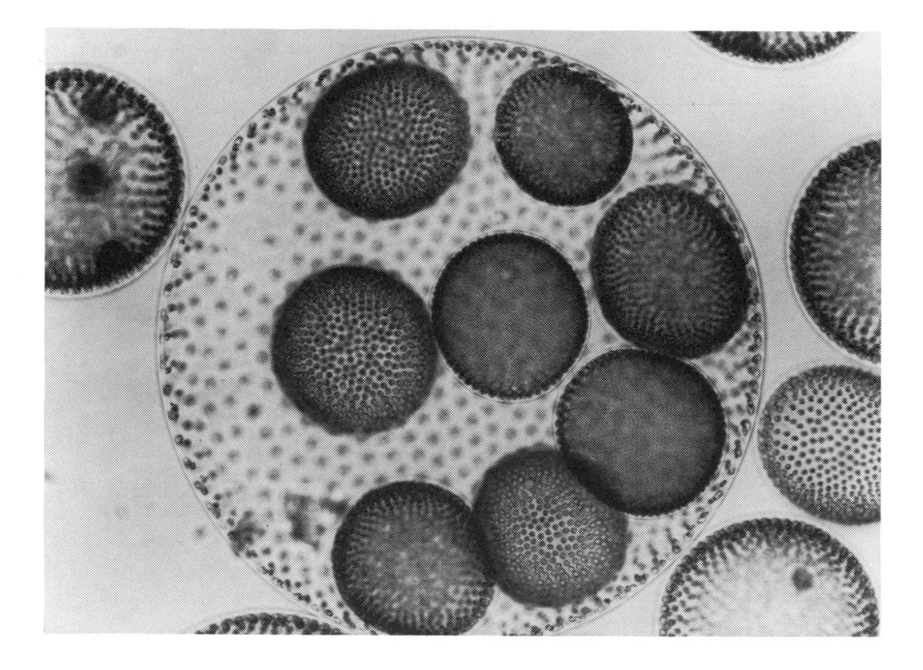

FIGURE 6-4

Volvox aureus **with daughter colonies.**

Photo courtesy: Carolina Biological Supply Company

To alleviate identification problems, you may wish to depend strictly upon biological supply houses for cultures. Refer to listing at the end of the chapter for some of the upcoming specimens; each of these has been called a colony and each offers a variation in the colonial existence.

Pandor-ina	This protozoan also forms a hollow sphere; it rests within a gelatinous envelope and is made up of approximately 16 biflagellated cells; it ranges from four to 32 cells in nature.
Eudor-ina	This colony is somewhat larger than *Pandorina;* it is composed of 32 spherical cells loosely arranged in a gelatinous sheath.
Gonium	*Gonium* is a plate-like colony composed of 12 to 16 interconnected biflagellated cells.
Platy-dornia	A horseshoe-shaped colony of 16 cells, *Platydorina* is flattened in appearance.
Steph-ano-sphaera	This is a circle of eight cells.
Volvul-ina	This is a spherical colony containing from four to 32 cells.
Meris-moped-ia	A flat sheet of cells, one layer thick.

All but *Merismopedia* and *Eudorina* can be maintained in a soil-water medium. You can prepare this medium by placing about one-half inch of garden soil in a test tube. Add distilled water until three-quarters full. Plug with cotton and steam for one hour on two consecutive days (Ward's Culture Leaflet No. 3, adapted from Pringsheim, 1946). A small pinch of calcium carbonate can be added prior to preparation to insure an alkaline condition that many algae prefer.

Do not throw away old cultures; there may be inactive spores or a dormancy condition that can later bring forth new life when activated. (Refer to Turtox Service Leaflet No. 6.)

When studying the various colonies you might consider a number of questions:

1. How many individuals make up each colony?
2. Is the fact that the colonies often reflect an obvious even number of cells (2, 4, 8, etc.) significant in terms of evolution? Why?

3. Are the individuals interconnected in any way? Do they function as one individual?
4. Do certain individuals appear to do specific jobs for the group as a whole?
5. How do the various colonies reproduce?
6. Is the word "colony" a word that always refers to a division of labor, or specific jobs being performed by certain individuals? How would you define the word "colony" after examining these various forms?
7. Compared to these examples, is *Volvox* a "more advanced" or "less advanced" colony?

While plant colonies often have obvious interesting movement, it is more difficult to detect movement in larger plants. The carnivorous (or insectivorous) Venus' flytrap, however, is able to move quickly and provide more than just a passing interest.

LOCATING AND INVESTIGATING CARNIVOROUS PLANTS

There are five basic carnivorous plants that you can bring into your classroom: Venus' flytraps, sundews, butterworts, pitcher plants, and bladderworts *(Utricularia).*

The Venus' flytraps are perhaps the best known of the carnivorous or insectivorous plants. Unfortunately, they grow naturally only in parts of North and South Carolina. While they can be ordered from most biological supply houses, unless you are prepared to investigate them thoroughly and maintain them, it is recommended that you delete them from your studies.

The Venus' flytrap (Figure 6-5), like other carnivorous plants, lives in mineral-deficient soils; it is thought to have adapted to insect capture in order to obtain needed minerals. It can make its own food through photosynthesis as do most green plants, but it requires continual moisture and high humidity.

Keep the flytraps in a terrarium previously lined with an inch of gravel, covered over with an inch of dried sphagnum moss. Cover further with several inches of moist, acid (mineral-deficient) soil. Plant the flytraps (also sundews and butterworts) on a well-drained slope, with living sphagnum moss filling the open spaces.

Keep moist but do not soak, and cover the container with a sheet of plastic or glass. Situate the container where it will receive

FIGURE 6-5

Venus' flytrap, a fascinating bog organism.

Photo by Stewart Harris; courtesy: Hampton Roads ETV Assoc.

---

direct sunlight for part of the day. If water builds up, use a pipette or dropper to remove the excess. A drain or outlet can be built in the terrarium at one corner; this could eliminate excess water without too much difficulty.

You might consider the following investigations with Venus' flytraps:

1. Place ants in the nearby vicinity of the traps. Can small ants make the traps close? How could an adaptation to eliminating smaller insects help insure the plants of adequate minerals?

2. What causes a trap to close? Search for "trigger hairs" (trichomes) inside the traps. How many are present? Using a pin, give one "trigger hair" a quick stroke. Does this close the trap? Touch the same trichome again. Does the trap close? Repeat with another open trap but this time touch two different trigger hairs, one stroke each. Does the trap close? How could the requirement of two trips help insure a trap's chances of capturing living specimens?
3. How long does it take a man-made closed trap to open again?
4. Place a small centipede or sow bug in the immediate vicinity of the traps. If it is captured, shine a strong light on the plant and observe the animal's struggles within the trap. How does its continued struggles affect the plant? How long will it continue to move about? How much time passes before the trap opens again and how do the remains differ from the original animal? Can the plant digest the hard protective parts of arthropods? How would the remains be removed in nature and how many victims will a trap capture before turning brown and dying? If a trap is cut off at the base after capturing one or two specimens, will another take its place?
5. How would you explain the ability of a flytrap to close quickly? It has been reported that small electrical disturbances take place while the trap is closing. Can you confirm or deny this? One theory suggests the closing is a rapid wilting action. If the trap has less water over a period of time, can it close as quickly? Another explanation suggests unequal amounts of growth as being responsible for creating a tension that results in the trap's quick response when touched. Can you confirm or deny this?
6. While other leaves in nature have projections on their edges, of what significance is the array of projections on the leaves of the flytraps? Of what value are the nectar glands?
7. How does a flytrap reproduce? Does it form bulbs? Does it have a flower and seeds?

The sundew grows nearly world-wide and can be maintained with the flytraps. Place the sundew in the vicinity of small fruit flies or ants. How does the plant capture insects? How does it digest their soft tissues? How long does digestion take place?

The butterwort is found throughout North America in bog terrain. It is considered to be one of the simplest of the carnivorous plants; its leaf glands secrete a sticky substance that holds small animals until they are killed and digested. You might observe the butterwort in action by placing small ants in the immediate vicinity of the plant.

While the pitcher plants, flytraps, sundews, and butterworts can all be maintained within the same terrarium, the bladderworts will need to be housed in an aquarium containing small fish or other aquatic life.

Bladderworts thrive in bog lakes in nature and the bladders of this type of plant are fascinating to observe in the classroom. Place a small amount of a bladderwort in a Petri dish of water and observe under the dissecting microscope. Add a few aquatic earthworms or roundworms and observe their capture. How long does digestion take place?

The carnivorous plants are extremely interesting to study because of their unusual adaptations as well as their rapid movement. From the standpoint of movement, you may also find the sensitive mimosa a plant worthy of classroom investigation.

THE SENSITIVE MIMOSA

The sensitive plant *Mimosa pudica* can be ordered year-round from biological supply houses. While you can maintain the plant in the soil that accompanies the shipment, keep it well watered and near direct sunlight.

You might attempt the following investigations with the mimosa:

1. Gently touch the leaves of a mimosa. How does the plant respond to touch? Can some leaves respond while others on the same plant do not? How many different ways do the plant parts respond when touched? Of what value is the response? How is the response made?
2. Give one plant half as much water as another. Does this affect the plant's response to touch? Does it alter the time necessary for the leaves to open again?
3. Place one mimosa in room temperatures with other plants in a range of temperatures from near freezing to 100°F. Do temperature extremes affect the plant's response to touch? Can certain temperatures alone cause the plant leaves to close?
4. Add various salts (sodium chloride, etc.) in small amounts in the water given periodically to the plants. (Do not repeat for long durations and be sure to transplant upon completion if the plants are adversely affected.) Do certain salts affect the response to touch? Do they affect the plants in other ways?

5. Blow on the plant leaves or use a fan with variable speeds. Can wind affect the plant? What wind speed is necessary for a response? (How would you determine wind speed?)

The mimosa plant can be grown from seeds as can many plants; however, when it comes to germinating seeds, while many seeds are excellent for classroom use you may wish to consider the unusual irridated seeds.

GROWING IRRIDATED SEEDS—WHAT WILL COME UP?

Irridated seeds provide a way to study the effects of high doses of radiation on life. They can be used effectively in genetic investigations or environment-oriented studies.

You can purchase tobacco, barley, marigold, oat, and squash seeds from a number of biological supply houses. These seeds have from 20,000 roetgens (r) to 4,000,000 r dosages.

You can plant the seeds in soil that may come with them or in garden soil. Observe the germination rates and the differences among seedlings. Do 20,000 r seeds give rise to the same numbers and types of plants as 50,000 r seeds?

Prepare slides from the root tips of irridated oat seedlings. Is chromosome damage present? (Compare to normal oat seedlings.) Is more damage present in higher dosages of radiation?

Attempt to inbreed unusual specimens and isolate distinct mutations. Are most plant mutations beneficial? This is an excellent area for student research.

A great deal of research can also be done with plants having short life cycles and limited food and space requirements. For these reasons you may wish to experiment with molds.

THE GROWTH AND CHARACTERISTICS OF BREAD MOLD

There are a number of molds that you can study in the classroom, including the common *Aspergillus, Penicillium, Phycomyces,* and *Sordaria,* but one of the most abundant and easily studied molds is the black bread mold, *Rhizopus* (Figure 6-6).

You can order *Rhizopus* as well as a number of molds from biological supply houses. However, you can obtain a bountiful supply of molds, especially *Rhizopus,* by lightly sprinkling water

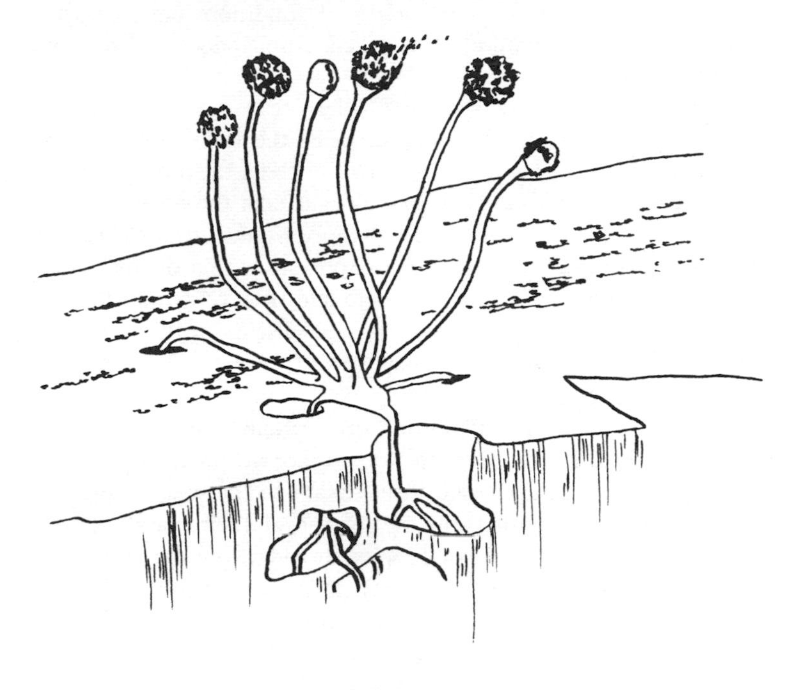

FIGURE 6-6

Black bread mold, *Rhizopus,* can be easily secured and maintained in the classroom.

---

on an old piece of bread and placing it in a plastic bag. Seal the bag and place in a warm dark area; observe periodically.

Bread, cake, cheese, oranges, bananas, and other foods can be used to cultivate molds, but for black bread mold, biscuits and various bread products are ideal. Breads lacking preservatives are preferred.

Rhizopus can be maintained on old bread with periodic transfers, or grown on nutrient agar.

You might wish to consider a number of investigations with molds, including the following:

1. Compare the growth of bread mold using a variety of breads. Do some breads support bread mold more readily than other brands? What types of bread best support the growth of black bread mold?

2. Place fresh mold cultures in a variety of conditions (varying light, temperature, and moisture). Which conditions does bread mold prefer for optimum growth?
3. Observe the cobweb-like growth of the bread mold (as seen on nutrient agar or on actual bread). Describe the various parts of the plant. Set the stage of a dissecting microscope on the outer edge of a fresh bread mold culture. How long does it take the bread mold to overrun the field of vision? Observe the various cottony filaments under a microscope. Is any movement apparent within the filaments?
4. Examine the spores of the bread mold and observe germination on nutrient agar. Describe the process and discuss why spores are necessary although the mold may spread vegetatively along the surface of the food supply.
5. Order plus and minus strains of *Rhizopus* and inoculate (as with bacteria) both strains on the same agar surface. Allow the two strains to grow toward each other and meet. Examine the sexual reproductive structures that result from the joining of their matlike fibers. The best "zygospores" or sexual spores should be present within one week from inoculation (or contact). After examination, dry the spores and save until ready to start new cultures.

Molds are plants that depend on living or once-living organisms or their products in order to exist. On the other hand, the larger green plants, utilizing photosynthesis, can make their own food; they provide the base of the food chains.

SELECTED EXPERIMENTS CONCERNING PHOTOSYNTHESIS

Photosynthesis is a complex process(es) that has been technically described in many texts; there have also been a number of technical experiments suggested in these texts. However, for some general understandings of photosynthesis and the plant as a whole, you may wish to consider some of the following ideas:

1. Place a piece of cardboard over green grass. After several days remove the cardboard and examine the grass beneath it. Compare the grass to the nearby uncovered area. What has happened and why? What happens to plants that never receive sunlight due to the place where their seeds germinated?
2. Obtain *Elodea* plants from a nearby aquarium shop. Observe a leaflet beneath the microscope. Where is the green color (chloro-

phyll) located in the cells? Observe and describe the motion of the chloroplasts. Does movement of the chlorophyll within the cell have any value? What causes the movement?

3. Peel the lower layer of cells carefully from the underside of a geranium leaf. Place on a slide and observe through a microscope. Are there openings on the underside of the leaf? What purpose(s) could these have in regards to photosynthesis? Place a drop of salt water on the cells and observe the change in the openings. Do the openings close? How could the closing of these openings (stomates) be valuable to a plant?
4. Place a plastic bag or bell jar over a geranium plant. Note droplets of water on the covering after a few hours. From where does the liquid originate? Place one plant in warm sunny conditions and another in cool darkness; observe the moisture build-up. Does temperature affect the rate of moisture loss (transpiration) from a plant?
5. Select two leaves from a living Coleus plant. Place Vaseline Petroleum Jelly on the lower surface of one leaf. Compare the two leaves over a long period of time. What happens and why? How is photosynthesis involved?
6. Compare venation in various leaves. How do the veins in a leaf relate to the potential effectiveness of photosynthesis?
7. Bring in leaves from a number of trees. Make tracings of their shapes, including pine needles as well as oak and maple leaves. Which leaves are best designed to capture the largest amount of available light? Do the largest leaves normally exist in direct sunlight or shade? Can pine needles exist in shade? How could leaf shape and situation relate to potential effectiveness of photosynthesis?
8. Place several of the leaves in front of a bright light source (slide or film projector) in a dark room. Which leaves allow light to penetrate through them? How could the thickness of a leaf affect the amount of photosynthesis that takes place?
9. Carefully boil spinach leaves for five to 10 minutes in a flask of ethyl alcohol that has been previously placed in a pan of boiling water. (*Do not* allow flames or a hot plate burner to get close to the flask with the alcohol.) After the chlorophyll has been removed, use the liquid for at least two studies:

   a. Pour a small amount of chlorophyll solution in a beaker. Cut filter paper in narrow strips and suspend them in the solution for several hours. Examine the papers. Are there several substances in a leaf? Do all of these deal with photosynthesis? Of what value is the yellow substance in a leaf? Is more than one type of chlorophyll present?

b. Pour a small amount of chlorophyll in a thin, flattened vessel. Darken the classroom except for one window opening (a filmstrip projector may substitute). Hold a prism in the sunlight and reflect the produced spectrum on the darkened wall. Pass the chlorophyll solution just behind the prism. Are certain colors absorbed by the chlorophyll? Which colors (wavelengths) are primarily used in photosynthesis?

10. Examine roots, stems, leaves, seeds, and other plant parts. Where is food from photosynthesis stored? How do animals obtain the food? Is all the stored (or converted) food available to animals in the food chain?

It is easy to forget that we are surrounded by life when driving down a road with only trees in the field of vision. We sometimes forget that tall trees are living plants; we are also guilty of neglecting the smaller plants that have their own fascinating adaptations for survival.

INVESTIGATING PRIMITIVE LAND PLANTS

Smaller plants have great value in the classroom. Not only do they often provide answers to problems in specific habitats, but they are frequently easy to maintain in the classroom year-round; their small size indicates their lack of demand of classroom space.

There are several kinds of comparatively simple or primitive land plants that you may wish to consider: lichens, mosses, liverworts, club mosses, ferns, and horsetails (Figure 6-7).

Living specimens of these plants can be purchased from biological supply houses. However, most of these plants can be found locally. You might wish to study the plants only in the natural habitat, but these can be easily maintained in the classroom. Simply prepare a terrarium similar to natural conditions. Keep moist, well drained, and humid.

The following questions are worthy of consideration when investigating these plants:

1. Where do these plants naturally exist? Which ones have fully adapted to dry land conditions?
2. How are these plants different from trees or larger flowering plants?

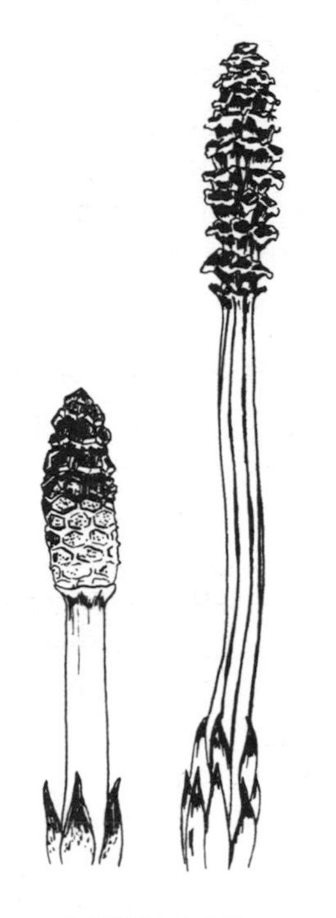

FIGURE 6-7

Horsetails are one example of primitive plants that can be brought into the classroom or investigated in the field.

---

3. When did these plants exist in largest numbers or where do these plants presently exist in largest numbers?
4. How do these plants reproduce?
5. How are mosses and liverworts different from ferns, club mosses, and horsetails?
6. How is each plant valuable and necessary in its specific habitat?

You may wish to germinate moss and fern spores on agar and observe the early stages of their life cycles. The germination kits are available from Carolina Biological Supply Company. All that is required is to gently dust spores on agar and observe over several weeks. In this way you can see a part of the life cycle(s) not normally seen.

You may also wish to collect some of the various kinds of lichens and determine if their algae can live without the fungus. What is their relationship?

While "primitive" land plants are interesting to consider from the standpoint of evolution, it is interesting to consider some of the recent discoveries of plant hormones and how we can accent development with them, giving rise to unusual specimens.

CONSIDERING PLANT HORMONES

Plant hormones are exciting. They can increase the size and amount of fruits and vegetables, induce early flowering, and influence root development. Gibberellin, for example, has enabled cabbages to grow six feet tall. In some cases growth hormones have been used as weed killers in our own back yards.

Some familiar plant hormones are gibberellic acid, kinetins, and auxins; four auxins are: indole-3 acetic acid or IAA, indole-butyric acid, napthalenaeacetic acid or NAA, and betanaphthoxyacetic acid or NoXA. However, there are also dwarfing compounds: N-dimethylaminosuccinamic acid-B-Nine and Phosfon, chemicals that produce abnormal cells (colchicine), as well as common weed killers that are powerful growth stimulants (2,4-D and 2,4-T). These chemicals are available from biological supply houses and, in the case of weed killers, they may be obtained from local garden centers.

You may consider a number of investigations with plant hormones, including the following:

1. Plant ten bean (or pea) seeds in two different pots of soil. Give both groups of seeds the same amount of water, light, etc., with the exception of giving one set of seedlings daily doses of dilute 2,4-D or 2,4-T. Compare the growth of the two sets of plants. Of what value could these chemicals be in regard to cultivating plants? How are they valuable in "killing" weeds?
2. Germinate a number of seeds in a Petri dish lined with several layers of moist filter paper; include pea, bean, radish, and corn seeds. In a similar container include similar seeds except in this container add dilute 2,4-D or 2,4-T. Compare root development of the two groups of seedlings. You may wish to repeat this, adding rancid butter with the seedlings; rancid butter is a source of IAA, an auxin; auxins are

also in 2,4-D and 2,4-T. What effects do auxins have on roots? How do they affect root development of a carrot top or an African violet leaf floating in water?

3. To determine the effect of auxins on stems, place young tomato or bean plants in boxes having only one source of light entry. Do the plants grow toward the opening, or light? How could this help plants to survive? Repeat with another set of plants, but this time spread rancid butter or NAA on the illuminated side of the stem. Can you make the stem actually bend *away* from the light? (Treat once or twice daily.) How do auxins enable plants to bend toward light? Why did the addition of auxins on the illuminated side of the stem cause the plant to grow away from light?
4. Purchase or grow bean, pea, tomato, or geranium plants. Treat parts of or entire plants with gibberellic acid on a cotton swab, every three days. Compare the treated plants or parts to a control group that has been grown in normal conditions. If you can treat green tomatoes or grapes you may be able to witness the value of gibberellic acid on increasing fruit size.
5. Plant dwarf peas in two different containers. Add gibberellic acid to one set. Can plant hormones override heredity? Examine the two sets of untreated plants for differences in height.

For additional experiments refer to Ward's Culture Leaflet No. 24, "Observations on Hormones and Plant Growth."

In summary, there are many exciting investigations that involve plants or plant-like organisms. This chapter has touched on only a few of the specimens you can examine; why not consider many more during the course of your school year?

BIBLIOGRAPHY AND REFERENCES

Baker, H.G. *Plants and Civilization.* Belmont, Calif.: Wadsworth Publishing Company, 1970.

Bold, Harold C. *The Plant Kingdom.* Englewood Cliffs, N.J.: Prentice-Hall, Inc., 1970.

Carolina Biological Supply Company Service Leaflets: "Bacteria and Fungi," "Carnivorous Plants," "Culturing Fern and Moss Spores," "Exercises in Bioluminescence," and "Living Ferns and Fern Allies." Burlington, N.C., Carolina Biological Supply Company.

Fagle, David L. "Bacteriology Course." *Biology Science Teaching Tips from TST (Science Teacher) 1960-1966.* Washington, D.C.: National Science Teachers Association, 1967, pp. 135-137.

Kotch, William J. "Fungi in the Laboratory." Chapel Hill, N.C.: University of North Carolina, 1968.

Poole, L. and Poole, G. *Insect-Eating Plants.* New York: Crowell Company, 1963.

Reid, Kenneth. "Project: How Many Insects Do Cobra Lilies Eat Each Year?" *Science World* (March 7, 1969), pp. 12-13.

Tortora, Gerald J., Donald R. Cicero, and Howard I. Parish. *Plant Form and Function: An Introduction to Plant Science.* New York: Macmillan Company, 1970.

Turtox Service Leaflets: No. 6, "Growing Fresh-Water Algae in the Laboratory"; No. 10, "The School Terrarium"; No. 25, "Non-Flowering Plants"; No. 32, "The Culture and Microscopy of Molds." Chicago: CCM: General Biological, Inc.

Umbreit, Wayne. "Should Microbiology Be a Part of Secondary Education?" *Biology Science Teaching Tips from TST (Science Teacher)* 1960-1966. Washington, D.C., National Science Teachers Association, 1967, pp. 126-128.

Ward's Culture Leaflets: No. 14, "Culture of Algae in the Laboratory"; No. 15, "Byrophytes and Pteridophytes in the Classroom"; No. 16, Rearing Insectivorous Plants"; No. 20, "Use of Terraria in the School Laboratory"; No. 21, "Germination Experiments"; No. 23, "Culture of Bacteria and Molds in the Classroom"; No. 24, "Observations on Hormones and Plant Growth." Ward's Natural Science Establishment, Inc., Rochester, N.Y.

Weier, T. Elliot, Ralph Stocking, and Michael G. Barbour. *Botany: An Introduction to Plant Biology.* New York: John Wiley and Sons, 1970.

AUDIO-VISUALS*

Films:

A Green Plant and Sunlight (EBEC)
Bacteria (EBEC)
Bacteriological Techniques (Thorne)
Carnivorous Plants (Moody Institute)
Energy and Its Transformation (EBEC)
Fungi (EBEC)
Gift of Green (Sugar Information, Inc.)
Leaves (EBEC)
Life of the Molds (Sterling Movies)
Microscopic Fungi (McGraw-Hill)
Movement in Plants (Coronet)
Photosynthesis (EBEC)

*Refer to glossary (page 211) for key to abbreviations and addresses; audio-visuals largely for upper elementary through high school students.

Films (cont.)

Photosynthesis: Chemistry of Food-making (Coronet)
Plant Growth (EBEC)
Plants Obtain Food (Indiana U.)
Reproduction in Plants (Coronet)
Roots (McGraw-Hill)
Roots of Plants (EBEC)
Seed Germination (EBEC)
Simple Plants–Algae and Fungi (Coronet)
Stems (McGraw-Hill)
The Growth of Plants (EBEC)
The Riddle of Photosynthesis (U.S. Atomic Energy)

Film Loops:

Air Plants (McGraw-Hill)
Budding of a Yeast Plant (McGraw-Hill)
Budding Yeast Cells (CCM)
Carnivorous Plants (Doubleday and McGraw-Hill)
Climbing Flowers (Doubleday)
Desert Plants (Doubleday)
Flowers Opening (Doubleday)
Fresh Water Algae (Doubleday)
Fruits Ripening (Doubleday)
Germination of a Seed (McGraw-Hill)
Growth and Pollination of Corn (Doubleday)
Growth of Molds (McGraw-Hill)
Growth of Mushrooms (McGraw-Hill)
Growth of Woody Plants (McGraw-Hill)
Mushrooms (Doubleday)
Penicillium (CCM)
Phototropism (McGraw-Hill)
Plant Behavior (6 loops) (Ealing)
Plant Reproduction (8 loops) (Ealing)
Plants (6 loops) (Ealing)
Production of Oxygen by Green Plants (McGraw-Hill)
Rhizopus (CCM)
Volvox (CCM)
Water Plants (McGraw-Hill)

Filmstrips:

Bacteria and Fungi (PSP)
Carnivorous Plants (Carolina Biological)
Classification of Plants (9 filmstrips) (BICO catalog)
Collecting Non-Green Plants (EBEC)
Fern Life Cycle (Carolina Biological)
Ferns and Horsetails (Long Film Service)
How Flowers Transfer Pollen (BICO Catalog)
Learning About Familiar Plants (BICO Catalog)
Liverworts and Mosses (Jam Handy)
Pine Life Cycle (Carolina Biological)
Plant Life (6 filmstrips) (BICO Catalog and EBEC)
Plant Study (4 filmstrips) (BICO Catalog)
Plants and Their Environment (4 filmstrips) (BICO Catalog)
Seed Plant Series (6 filmstrips) (IFB)

LIVING PLANTS OR PLANT-LIKE ORGANISMS

	BICO	*Carolina*	*Conn.*	*CCM-Turtox*	*So. Bio.*	*Stansi*	*Schettle*	*Stein*	*Wards*
Bacteria: (most species)	3.75	4.00	4.00	4.00	–	4.00	5.00	4.25	4.25
Allomyces	–	4.00	–	–	–	–	–	–	–
Volvox	2.00	2.50	2.50	2.50	2.80	2.50	3.00	2.75	4.00
Pandorina	–	3.50	2.25	2.50	2.65	2.25	–	2.75	4.00
Eudorina	2.00	3.50	2.25	2.50	2.65	–	3.00	2.75	4.00
Gonium	2.00	3.50	2.25	–	2.65	–	3.00	2.75	4.00
Venus' flytrap	1.75	2.00	1.75	2.60	–	2.50	2.60	1.50	2.50
Sundew (ea.)	1.25	1.00	1.00	1.55	–	1.50	2.80	–	1.25
Pitcher plant	2.25	2.50	1.75	2.85	–	3.00	3.15	–	2.50
Butterwort	–	2.00	–	–	–	–	3.95	–	–
Bladderwort (*Utribularia*)	–	3.00	3.50	4.00	–	3.00	–	–	3.00
Mimosa (sensitive plant)	–	1.25	–	–	–	1.25	–	–	–
Irridated seeds	–	4.50	–	4.60	–	1.50	–	–	–
Molds (*Rhizopus, Aspergillus, Penicillium, Phycomyces,* and *Sordaria*)	3.75	4.00	4.00	4.00	–	4.00	5.00	4.25	4.25
Rhizopus (+ and –)		6.50	6.50	8.00	–	8.00	10.00	8.50	8.50
Moss and/or fern spores germination kits.	–	2.50	–	1.75	–	4.00	–	–	
Liverworts	1.10	1.25	2.00	2.60	–	2.25	4.00	–	1.50
Mosses	.90	1.00	1.00	2.05	–	1.00	1.75	–	1.50
Ferns	1.00	1.25	1.00	2.06	–	3.00	1.70	–	2.00
Horsetails	–	–	–	–	–	–	3.40	–	3.00
Club mosses	–	1.30	2.50	2.20	–	3.00	3.40	–	3.00
Plant hormones:									
Gibberellic Acid	2.50	3.00	3.00	2.80	–	2.00	2.95	–	1.00
Auxins (1 oz)	1.35	1.25	1.50	2.60	–	23.00*	2.80	–	2.50

*kit of 17 hormones

Armstrong Associates, Inc., Box 127, Basking Ridge, N.J. 07920 offers sundews, 6 for $5.00; flytraps, 6 bulbs for $3.00; butterworts, 3 for $3.00, and a host of pitcher plants: Parrot pitcher, hooded pitcher, miniature huntsman's horns, huntsman's horns, northern pitcher, and cobra lily, each 3 plants for $3.00 except tor the cobra lily which is 1 plant for $2.50.

Edmund Scientific Co., 615 Edscorp Building, Barrington, N.J. 08007 offers 3 Venus's flytrap bulbs for $2.00 and an assortment of a sundew, purple pitcher plant, and a small huntsman's horn for $4.50 postpaid.

All of the prices listed were in effect at the time of writing; request current catalogs for present listings and prices.

SEVEN

Investigating Vertebrates

If you ask students which animals they would prefer to study, the majority of responses would probably favor vertebrates: fish, amphibians, reptiles, birds, and mammals. Vertebrates are larger, more familiar animals. They are our favorite pets; they are our food; they are ourselves.

Vertebrates are excellent for classroom study. They are easily obtainable, they can survive well, and they exemplify features and behavior of the larger forms of life. Nevertheless, there are disadvantages.

DIFFICULTY AND PROBLEMS INVOLVING VERTEBRATES IN THE CLASSROOM

Far too often vertebrates are brought into the classroom and abused through carelessness. While an insect, flatworm, or protozoan need not be fed regularly, a mammal, fish, or other vertebrate may require one or more daily feedings for sustained survival.

While a smaller invertebrate may require little more than a dish of water, vertebrates require extensive classroom space and a wide range of foods and special care if they are to be humanely treated.

While diseases do not seem to be a serious problem among smaller invertebrates, larger vertebrates may carry bacterial and virus infections that at times may be transmitted to man.

When working with invertebrates there is little concern if a protozoan culture evaporates, thereby killing thousands of individuals; their numbers can often be replenished through the inactive cysts left behind. If a large vertebrate dies through carelessness, the odors and difficulty of disposal are but small problems. A more serious issue is the example it provides concerning treatment of animals. It could reflect your concern for life itself.

While many vertebrates are not housed or fed properly through ignorance, Walker (1955) also notes that some individuals may wish harm to animals. The matter becomes complex when classroom investigations require extensive manipulation of vertebrates. Orlans (1968) suggests that "animal experimentation involving surgery should be confined to research institutions and hospitals with adequate equipment and proper supervision."

So, you may wish to investigate living vertebrates, but only those aspects that do not threaten or demand the taking of a life.

SELECTED STUDIES WITH VERTEBRATES

Amphibians

There are three basic amphibians to be considered: frogs (Figure 7-1), toads, and salamanders. While biological supply houses supply a variety of these amphibians, when possible you may wish to obtain amphibians locally. If necessary, purchase only those amphibians that can be returned to local habitats upon completing investigations.

Keep frogs and toads in moist, but not wet, terrariums. A shallow pool of fresh water may be kept at one end although this is not necessary. Provide a few sections of logs and rocks for hiding purposes, and add sphagnum moss, covering the entire floor of the terrarium. You may also add weeds, ferns, or woodland mosses; these help hold moisture and humidity while utilizing frog wastes.

Keep them in indirect sunlight or expose them only periodically to direct sunlight; the air temperature should be between 65-80°F. Cover the container with a sheet of plastic having a few air holes.

Periodically feed them *live* earthworms, mealworms, cock-

FIGURE 7-1

In the habitat, the elevated eyes of the grass frog enable it to view above the water's surface while remaining in large part out of sight.

Photo by Stewart Harris; courtesy: Hampton Roads ETV Assoc.

roaches, or small crickets. They seem to feed more readily when in groups; younger specimens also seem to feed more readily than older, larger ones. Keeping fewer specimens not only insures better chances of sustained survival, but will require less food.

If frogs and toads are both available, prepare two terrariums; do not mix frogs and toads in the same terrarium.

If frogs refuse to feed and begin to develop a reddish color on their legs, scrub them immediately in salt water; this may prevent a red-leg fungus from killing them. You may discourage fungus growth (in general) by placing a few copper pennies on the moist ground/moss of the terrarium; the iodine in sphagnum moss may discourage red-leg fungus from becoming a problem. If any frog or toad becomes sick, immediately isolate it from the remaining specimens.

Tadpoles can be maintained in fingerbowls or in aquariums containing charcoal-filtered pond water. Place in sunlight for at least a few hours daily and occasionally add small bits of boiled spinach or lettuce; you might also add fresh-water algae that has been field collected.

Salamanders (including newts, Congo eels, and sirens) may be fed live earthworms, mealworms, *Daphnia,* or bits of fish. The larger specimens may be offered small minnows and young frogs. Keep in aquariums with large rocks having inclined surfaces that enable the salamanders to leave the water. Be sure to cover the aquariums.

Some of the studies with amphibians may include the following:

1. Order fertilized frog or toad eggs from biological supply houses and observe development into adults.
   a. Of what value is the jelly coating of the eggs?
   b. Describe the tadpole's early development within the egg masses.
   c. What do tadpoles eat?
   d. Does the tadpole drop off its tail when it becomes an adult? What happens to the tail?
   e. How long does a tadpole remain in this stage?
   f. When are tadpoles found in the natural habitat?
   g. What eats tadpoles in the natural habitat?
2. Obtain a grass frog.
   a. Count the number of pulses its throat makes in ten seconds and multiply by six. How many gulps of air does it take in one minute? Do colder or warmer temperatures, bright lights, or the presence of people or loud sounds affect the breathing rate?
   b. Carefully spread the webbing of a frog's hind foot under a microscope. Describe the blood cells, the vessels, and blood flow inside the frog.
   c. Does the frog have an ear? Can you devise a way to find out how well it hears?
   d. Offer the frog live mealworms and earthworms. Describe its feeding behavior.
   e. Measure the lengths of the forelegs and hindlegs. How are their lengths and functions different? How far does your "average" frog jump? Hold a frog-jumping contest and find the answer.
3. Obtain a toad. Compare and contrast the toad and the frog. A toad's

skin produces an irritating substance that discourages predators from eating it, but does it produce warts on humans? Experiment with student volunteers and find the answer.

4. Place a dish of water near a toad. Does it ever use its mouth or tongue to drink? Why is a good way to give a toad (or frog) a drink of water simply to pour water on its skin?
5. Obtain a tree frog or toad. Of what significance are the suction discs and coloration? Why does it also have to return to water as do other amphibians?
6. Obtain a salamander or newt. Observe its feeding behavior; offer it live sections of earthworms or brine shrimp. Describe coloration and other adaptations that enable salamanders to live in their natural habitat. Observe their eggs in season.
7. Obtain an amphiuma or Congo eel. (Be careful when handling this large salamander.) How do the small legs of the amphiuma enable it to survive in a habitat filled with sticks and debris (Southern swamps and ditches)? How does its slender appearance also help in movement through narrow places? Are its eyes of much value in a muddy environment? Observe its feeding behavior; provide living earthworms in its aquarium.
8. Obtain a mud puppy (*Necturus*—see Figure 7-2). Compare and

FIGURE 7-2

The mud puppy (*Necturus*) has a cluster of gills behind its head; the movement of the gills indicates the amount of dissolved oxygen in the water.

Photo by Stewart Harris; courtesy: Hampton Roads ETV Assoc.

contrast this salamander to others you have studied. How do the movements of the gills indicate the amount of oxygen in the water? From what behavior or feature does this salamander receive its name?

For further information on frogs and toads you might refer to Turtox Service Leaflets No.7, "The Care of Frogs and Other Amphibians," and No.23, "Feeding Aquarium and Terrarium Animals"; also, Carolina Biological Supply Company leaflets: "Culture of Amphibian Larvae," "Care and Feeding of Amphibians and Small Reptiles," and "Terrarium Care."

Fish

You may wish to keep both marine and fresh-water fish. Refer to Chapter Five for details on establishing a salt-water tank.

You can establish a fresh-water aquarium by thoroughly cleaning a tank and covering the bottom with one-half inch of garden soil. Cover with an additional two to three inches of *clean* sand and/or gravel. Slowly add pond or spring water, preventing direct contact with the sand. (Tap water can be used, but add a dechlorinating compound and allow it to settle for at least 24 to 48 hours.)

Next, you should establish rooted plants such as *Vallisneria, Sagittaria, Elodea,* or *Ludwigia.* These and others (*Cobomba, Myriophyllum, Ceratophyllum, Lemna,* etc.) can usually be purchased from local aquarium shops although they are also available from biological supply houses.

As a rule of thumb you may begin by adding one inch of fish per gallon of water. Stock with animals that are compatible and include pond or lake species that normally exist in standing water. Do not overcrowd.

Keep the tank about two-thirds filled with water and cover with a glass plate or commercially available metal cover. Replace evaporated water with additional pond water. Keep near a window that will allow no more than two hours of direct sunlight daily. Feed commercial fish foods, *Tubifex* worms, and live water fleas (*Daphnia*) to your specimens, and attempt to maintain water temperatures between 40-60°F.

Your fish studies may include the following:

1. Obtain a goldfish, mosquito fish or substitute and place in a small aquarium. Observe gill covers opening and closing. What does this indicate? Record the water temperature and count the number of times the gill covers open and close in one minute. *Slowly* add crushed ice to the aquarium, observing and recording the number of gill cover movements at various temperatures approaching 32°F. Slowly add warm water to the aquarium. Again record the number of gill cover movements at various temperatures approaching 85°F. Graph the results and answer the following questions:
   a. How do colder temperatures affect the fish?
   b. How do warmer temperatures affect the fish?
   c. At what temperature(s) does the fish require the most oxygen?
   d. At what temperature(s) does the fish require limited oxygen?
   e. Is the water temperature also the internal temperature of the fish?
   f. Do younger or smaller fish respond to temperature changes differently than older or larger fish?
2. Purchase eggs of the Japanese Medaka (Figure 7-3) from the Carolina

FIGURE 7-3

The female Japanese Medaka provides eggs that can be studied through their development; they offer a spectacular view of a developing circulatory system.

Photo courtesy: Carolina Biological Supply Company

Biological Supply Company or obtain eggs locally from angel fish or other egg layers. View them under the microscope as they develop. You should be able to observe the spectacular development of a fish's heart and circulatory system. (The Japanese Medaka development is described in detail in a booklet offered by Carolina Biological; it would be wise to purchase the booklet with the eggs.)

3. Place a mixture of tropical fish in an aquarium. Observe the various places the fish congregate. Do certain fish prefer certain areas? Do some fish attack others? Which fish are compatible? (Remove bullies.) Will a "bully" fish become less aggressive if put in with larger or more aggressive fish?
4. Allow a mated pair of stickleback fish to occupy an aquarium. After a few weeks introduce a new male. Describe the behavior that follows. Do fish hold or defend territories?
5. Obtain mosquito fish (*Gambusia*). Start the fish in fresh water and *slowly,* over a period of weeks, add salt water to the tank. Can this fish adapt to brackish or marine conditions? Repeat with minnows, perch, croakers, catfish, etc. How could a fish's ability to live in both salt and fresh water be an advantage to survival?
6. Build a maze within an aquarium; introduce goldfish and, using bright lights as indicating a wrong turn, see if you can "teach" a fish to swim a desired path, reaching a certain chamber where it is fed. Do fish appear to be able to learn?
7. Feed goldfish or guppies at the same corner of an aquarium each day. Do fish appear to learn when and where to be fed? How long does it take them to "learn" a feeding location? Can you use a signal (light or sound) to attract them to the surface? (Initiate the signal when feeding and then use it by itself; can the signal alone bring them to the surface?)
8. Purchase guppies and attempt to breed them. Does a female guppy give birth to the same number of offspring each time? Will the parents eat the offspring? How would you attempt to breed fish, emphasizing a desirable trait?
9. Offer a mated pair of fish an excess amount of food. (Remove uneaten portions after feeding, or change the water frequently.) Limit the food given to another set of similar fish, of the same age if possible. Does limiting food affect eventual growth, size of offspring, or length of life?
10. Keep one set of young goldfish or guppies in cold water (35-40°F.) and another set in warmer temperatures (80-85°F.). Compare behavior, reproduction rates, growth, and life span. How does temperature affect fish in various latitudes or under exceptionally hot or cold seasons?

11. Obtain mosquito fish and keep in various daylengths. Can the amount of daylength affect reproduction or other processes?
12. Provide one aquarium with an excess of air stones or aerating devices, constantly bubbling air through the water. Provide another aquarium with only the water/atmosphere boundary as a source of oxygen. Stock both with the same kind and age of fish. Does an increase of oxygen affect fish?

You may wish to investigate other scaly vertebrates during your school year; this could include reptiles.

Reptiles

Reptiles are not the best vertebrates to bring into a classroom. They are often sluggish; many prefer to hide and some refuse to eat seasonally; they are rarely friendly, and of course, some may bite.

It is suggested that you take great care in handling reptiles. It is hoped you will secure local reptiles for brief studies and then return them to the natural habitat.

If local reptiles are not available or obtainable, you can purchase three types of reptiles from biological supply houses: lizards, turtles, and snakes. If possible, secure specimens that can later be released safely into your local habitats.

Note: Crocodiles (Caimans) can be purchased from local pet shops, but this is discouraged if only because future release is impossible. Great caution is also suggested when investigating snakes.

Maintaining Reptiles

Keep your lizards (see Figure 7-4) in a warm, covered, woodland terrarium. A rotten section of log or hollowed-out wood may provide a home. Small branches and leaves may also provide cover for the lizards. You can feed lizards live mealworms, flies, or small earthworms. Occasionally offer them lettuce or small pieces of ripe bananas. While making a small container of water available, it is also suggested that you sprinkle water daily on the vegetation in the aquarium (and on them occasionally). If you purchase any lizards, request medium-size or younger specimens; they are generally more active and feed more readily. Hold lizards by using your thumb and forefinger, applying pressure on a front leg while

FIGURE 7-4

The fence lizard (*Sceloporus*) is one species offered commercially.

Photo by Stewart Harris; courtesy: Hampton Roads ETV Assoc.

---

using your hand to gently support the body; do not pick up the lizard unless absolutely necessary. (Avoid grabbing a lizard by the tail.)

Land turtles should be fed earthworms, lettuce, apples, tomatoes, or strawberries. They may also eat raw hamburger. Land turtles may not feed during the winter months, when they would normally be in hibernation. In addition, land turtles may not be active during the winter months, even in a warm classroom. Keep them in a woodland terrarium having a large amount of dried leaves and debris where they can conceal themselves. Be especially gentle when handling turtles, supporting them firmly when lifting them. Never drop them; a cracked shell could kill them. Do not keep them in the classroom for more than a month and release them as soon as possible into their natural environment.

Fresh-water or pond turtles can be kept in shallow aquariums having rocks or an exposed area where the turtle can leave the water periodically. A floating section of cork makes an excellent floating platform. (Make certain the turtles can leave the water voluntarily; otherwise they may drown or become sickly.) Feed your pond turtles strips of fresh or frozen fish, live earthworms, live brine shrimp, ant eggs, or commercial food pellets; occasionally offer them lettuce and fruit. Remove excess food directly after feeding, and change the water when it becomes cloudy or develops an odor. Keep away from direct sunlight except for short durations; even so, give them an option for shade if direct sunlight is available.

Note: In general, most vertebrates, especially reptiles, seem to thrive best when they receive natural sunlight directly or indirectly for a part of each day.

Snakes demand a great deal of space and although conditions vary according to species, they do well in a woodland terrarium. Dry leaves, twigs, stones, and debris make excellent hiding places for many non-poisonous snakes. They also provide sharp materials to push against when shedding their skins. Offer your non-poisonous snakes a variety of live foods, including earthworms, slugs, mealworms, crickets, and other insects. Newborn mice are preferred by some species and are essential for rat snakes and constrictors. (Feeding live mice or other prey to snakes may be difficult to accept, but these reptiles have adapted to *live* prey. While they may go for extended periods of time without feeding, it is essential they be offered live food; slow starvation is extremely cruel).

If a snake is captured locally, it is recommended that you hold the animal for only a few days before releasing it. In this case feeding is not essential, but sprinkle water in the containers and provide a clean dish of water daily. Also provide an outlet for fresh air. Handle snakes with great care, grasping them firmly behind the head, supporting the body with the other hand. Use gloves as a precaution; do not drop the snakes or bend them about carelessly. (If an excess of mites are seen in the container, immediately transfer the snake to another cage, removing any mites from its body.)

Your activities with reptiles might include the following:

1. Obtain a "chameleon" or anoli lizard (Figure 7-5); these are available commercially.

FIGURE 7-5

The American chameleon (*anoli*) is a popular lizard for classroom investigation.

Photo courtesy: Carolina Biological Supply Company

---

a. Offer the lizard live mealworms. Observe the feeding process. Is there more than one way that a lizard can ingest food? Offer the lizard various foods; which does it prefer? How would a desert lizard obtain both food and moisture from a single feeding?
b. Observe the movement of a chameleon's eyes. How are they different from your own? How does their movement help this vertebrate to survive?
c. Place an anoli lizard in direct sunlight. Situate an insulated strip directly next to a portion of its body. Later remove the strip and compare the color of the entire animal to the previously covered portion. What causes a lizard to change color? How could it help in survival?
d. Of what value is the lizard's tail? Although it is not recommended that you attempt this, you may discuss the lizard's ability to release its tail. How could this help with survival?

2. Obtain a pond turtle from a biological supply company or from local pet shops.

a. Offer the turtle live earthworms or a slice of fresh or frozen fish. How does the turtle use its beak and feet when feeding? How is the turtle of value in the natural habitat through its food preference?
b. Carefully examine the turtle's hind feet. Of what value is the webbing? Do land turtles have webbing on their hind feet?
c. Observe the turtle's shell design and coloration. How do these facilitate swimming or other means of protection/survival?

3. Obtain a land turtle (Figure 7-6) from a biological supply company or a local pet store (or from the natural environment).

FIGURE 7-6

The Cumberland Terrapin.

Copyright: Turtox

---

a. Compare and contrast the shell design, feet, and other adaptations of the land turtle to the pond turtle. How is each suited to its specific environment?
b. Draw a two-foot (radius) circle around a turtle; how long does it take the turtle to crawl two feet? How fast does a turtle move in miles per hour? (5,280 ft. per mile.)
c. Make a small painted mark on the shell of a turtle and release in the natural habitat. Follow its immediate movements and mark

them on a map of the area. If possible, determine if turtles move about much in their lifetimes.

4. Obtain a garter snake from biological supply houses or from local pet shops.
   a. Offer the snake live earthworms. Describe the feeding process. How is it possible for the snake to open its mouth so wide? What other foods will it eat? What is one of its values in the natural habitat?
   b. Carefully hold the snake. Is it slimy or smooth? Does it feel cold or warm to you? If given a choice of sunlight or shade in its terrarium, which will it prefer? Why? What is meant when you say a snake is cold-blooded?
   c. Examine the underside of the snake. Are the bottom scales identical for the full length of the animal? Can you see what is meant when it is said the non-poisonous snakes in North America have a double row of scales running from the anal opening to the tip of the tail? How do the bottom scales aid in movement?
   d. Observe the action of the snake's tongue? Does it dart out more frequently in unfamiliar or familiar terrain? How does it help the snake to sense the surroundings?

When returning your reptiles back into the natural habitat, you will probably become aware of the presence of a vertebrate that supposedly descended from reptiles—birds.

Birds

While you can study birds in the classroom, using canaries, parakeets, or pigeons, you may find your most satisfying studies in the field. (See Figure 7-7).

To begin, select a bird that you can easily identify; it should also be extremely abundant in your community. You might spend morning and afternoon hours observing this one species or, in particular, one or two birds. See if you can answer some of the following questions:

1. Does the bird live in a flock that remains together through the day? Does the flock roost together in the evening? Does it flock only seasonally? Does it flock only at time of feeding? What are the advantages and disadvantages of flocking?
2. Does the flock have leaders? (Observe crows.) Does the entire flock move when a certain individual(s) does? Will the group move on signal from the leader(s)?

FIGURE 7-7

Birds, including the bluejay, are best studied in the field.

Photo by Stewart Harris

---

3. If the bird you study is not a member of a flock, what are its habits and behavior?
4. Many birds defend territories. Does your species defend its territory? Which sex defends the territory and against whom? Of what significance is the bird's song or the male's bright plummage in defending a territory?
5. In your field studies, some of your goals may be to:
   a. list as many identifiable species as possible;
   b. make vertical maps showing birds that exist at different levels in the habitat;
   c. make charts showing the most numerous species in your community and where they exist;
   d. identify which species are compatible;
   e. identify which species possess "mobbing" characteristics, where groups will attack a much larger predator and drive it away;

f. list food preferences for various birds. How do their bill shapes and sizes indicate the type of food they prefer?

You may wish to construct a bird feeder to facilitate your observations. A two-foot square section of wood, nailed to the top of a long pole, provides one example of a bird feeding platform. Place various seeds, bread, or other foods such as fruits and nuts on the platform; observe and record the various types of birds that visit. Alternate foods; one day place only bread on the platform. Another day add only bird seed; another day only fruit. Do certain birds prefer only certain foods? (In summer you may also put out a hummingbird feeder filled with sugar-water.)

For background information you may wish to contact your local bird clubs or the Audubon Society. For additional information sources, refer to the bibliography at the end of the chapter.

If you have the opportunity to observe a bird nest with freshly hatched young, you will discover the parental attention that is typical only of birds and mammals among all the types of life in existence.

Mammals

Mammals are among the most frequent vertebrates we come in contact with; they are our pets—the cats, dogs, horses, hamsters, rabbits, mice, and gerbils. They are also our food: pigs, cows, and sheep. They are our clothes: wool and leather; and they are often our favorite animals at the zoo, perhaps because of their high level of intelligence or resemblance to man.

Cages for mammals depend on the animals in question. Small mice, rats, guinea pigs, hamsters, or gerbils should have as much room as possible, with shredded paper, wood chips, or dry leaves that provide nest materials. (A pair of mice in a 20-gallon aquarium is an excellent space ratio while anything additional to commercial cages is beneficial.) Rabbits should have from 8-15 square feet of cage space, depending on the size of the rabbit. (An exercise wheel may help if less space is available when keeping smaller mammals.)

A water bottle feeder should be provided along one side of the cage; keep it filled with fresh water. Feed small mice, rats, hamsters, or other rodents dry oats, peanuts, sunflower seeds, lettuce, carrots, pecans, hickory nuts, acorns, or other nuts. Commerical food pellets may supplement and/or substitute for other foods.

If young mammals are housed, you may wish to feed them with a doll's baby bottle if the mother is not available. Homogenized cow's milk, diluted with half as much water, may be used in the doll bottles. Be careful that you don't make too large a hole in the nipple or feed the animal too fast; this could choke a young mammal.

Since there are many domesticated mammals you can bring into the classroom, it is recommended that wild mammals be left in the natural habitat. Skunks, armidillos, raccoons, foxes, opossums, monkeys, large cats, or other mammals that appear in pet shops from time to time should be avoided. Do not encourage the capture of wild animals by purchasing them. Remember that many such mammals can be studied in the outdoor environment; squirrels, for example, can be fed and studied in city parks or even in your own backyard.

Of the various studies you can conduct with mammals, you may wish to discourage the nutritional experiments that have been done with rats. Their results are well known. On the other hand you can have a great many interesting studies with some of the following:

1. Allow rats to run a maze that you have built. Can a young rat solve the maze quicker than an old rat? Do groups perform better than individuals? Do parents teach their young how to run the maze?
2. Bring in dogs (or cats) and have students show their training results. Do mammals appear to learn more quickly than amphibians or reptiles?
3. Have students observe a mouse or other mammal and list the features that typify this group.
4. Observe the mouths of horses, dogs, rats, cats, and other mammals. How are a mammal's teeth designed to eat specific foods?
5. Conduct a pet show featuring mammals.
6. Keep a pair of mated mammals in your classroom for the duration of the year; much can be learned from continual encounters with mammals.
7. Visit a zoo and observe mammals that are not native to the local area. Compare and contrast to local mammals.

While you can study many types of mammals in the classroom, at the zoo, or on the farm, you can study one of the most important mammals by looking at yourself.

MAN, THE ORGANISM

We often forget that man shares many things in common with the life around him. He respires (breathes); he ingests, digests, and assimilates food; he excretes waste products; he responds to light, heat, touch, and other stimuli; he grows; he reproduces; he dies.

Man shares his greatest similarities with the vertebrates more specifically with mammals, most specifically with primates. Yet man can see his reflection in all life.

You may wish to discuss and investigate some of the following aspects of man, the organism:

1. Let students make their own fingerprints, using an ink pad. Compare and contrast fingerprints. Are fingerprints the only features that differ between humans?
2. Have students each place a section of cardboard in front of one eye. When facing direct sunlight or a light bulb (not too close), describe the pupils of the two eyes. Does the pupil of the covered eye act independently of the other eye? Of what value is the pupil?
3. Have students exhale one breath into previously stretched balloons. Compare balloon sizes and lung capacities. Quantitative figures can be obtained through displacing the balloons in a known amount of water or exhaling a breath through rubber tubing into a beaker filled with water and inverted into a tray of water. Of what significance is lung capacity?
4. (a) Use a toothpick to scrape skin cells from the roof of your mouth or between your teeth; you can also scrape skin cells effectively from the soles of your feet. Examine the cells beneath the microscope and calculate the number of cells in a square centimeter of skin.
   (b) Touch the point of a needle (gently) to a square centimeter of skin on your arm, recording any places that do not feel "touch" or pressure; repeat with a warm and a cold needle.
   (c) Count the number of hairs in a square centimeter of skin on your arm, and examine a hair beneath the microscope.

[From the preceding activities, describe the skin of a human.]

5. Hold your thumb directly upwards. Without using it, attempt to pick up objects, write, comb your hair, and tie your shoes. How does an opposing thumb help man to extend his capabilities beyond other primates?
6. List various drugs to which humans can become addicted. Where are they found in nature? How do they affect man? Can humans become addicted to alcohol, cigarettes, or even food? Can addictions be "cured"? What can be done? (Your local police may provide a speaker on the topic of drugs.)
7. Discuss the social aspects of humans. Why are laws necessary? Why must some humans be elevated in status above others? Why and how did communities and cities develop?
8. Discuss man's foremost asset: intelligence. Give evidences that humans can interpret and predict from observations.
9. List and discuss the various races of humans. How are they physically different and alike one another? Can you find any evidences of differences in intelligence?

MAN: OVERPOPULATION

While many organisms produce thousands of potential offspring per individual, man is normally limited to one individual per reproductive female per nine to ten months' time. Yet, while those organisms with thousands of offspring lose large numbers to environmental factors, man has a greater chance that his offspring will reach maturity.

In their habitats, those organisms that overpopulate pay the price; as with bacteria and others, the result of too many individuals means a loss of food supply and space as well as a build-up in waste products and other unfavorable conditions limiting the expansion of the species. Population declines generally follow peak densities. Where is man in terms of population growth? How high will our numbers reach before a decline? Some research has indicated that by the year 2020 to 2040, man will drastically decline in numbers. If this particular prediction model is correct, there is little that man can do to avoid a population crash. However, if the human population is lower in the future, there will be a less drastic crash. There are two major questions involved with the future: Can man limit his population? Can man limit his energy demands?

You might wish to include some of the following ideas in

your search to answer the first question concerning population expansion:

1. Make a graph of the world's human population since the construction of the Egyptian pyramids. When did the human population show the greatest gains in the past? What caused declines in numbers of humans in the past? From this evidence, what will *probably* limit our population expansion in the future? (Jacobson, 1972.)
2. Discuss the advantages of birth control methods. You might want to include questions such as: "what about those individuals that cannot use present methods of birth control, or simply will not"? "How safe are prescribed means of birth control?" "Can having only two children per couple hold the population rate if the children reproduce another generation before their parents are in their 40's"?
3. Discuss the impact of one group limiting its population. If only the "upper class" can afford to have children, what will become of the present "lower class"? If few offspring were possible today, would these few offspring later be willing to support a massive older population? If any one group in our society does not contribute to the gene pool, what will eventually become of their traits or characteristics?
4. Have local doctors and other authorities discuss the issue of "abortion on demand" in a panel discussion.
5. Make a "family tree" with two imaginary couples that each have five children. If their children intermarry, and each couple, has another five children (within 25 years from the original couple's marriage), how many humans would result if this continued for 100 years?

ENERGY REQUIREMENTS

At present, two major aspects of energy seriously concern humans: the food energy necessary for existence, and the energy necessary for our industrial way of life. Both are threatened if the human population continues to reproduce and live as it has in recent years.

You may wish to consider some of the following areas in your classroom discussions of energy requirements:

Food

1. Display vegetables, meats, fruits, milk, and bread. Which foods are in current shortage? What affects the price of a food item? Why are

certain foods not as available as others? Why do foods high in protein often cost more?

2. Have your school nurse or cafeteria manager talk to your class on nutrition.
3. What parts of the world presently lack an adequate food supply? Are some people in the United States suffering from starvation? Where and why?
4. What are the causes of some people storing an excess amount of energy (in the form of fat)? What are the social aspects of being fat? Are there places in the world where being "fat" is a status symbol?
5. Grow various vegetables in the classroom. What are some of the problems involved with raising crops? What is being done to increase the yields? Invite local farmers or agriculture agents to discuss what is being grown locally.
6. Discuss "farming the sea" and other futuristic hopes of increasing the amount of available food.

Power

1. Display fossil fuels. Where are they found? How did they form? How abundant are they? What are some of the problems involved with utilizing their energy?
2. Visit an atomic energy plant visitor center. What are the advantages and disadvantages of atomic energy?
3. Purchase a solar cell or oven from Edmund Scientific Company, 615 Edscorp Building, Barrington, N.J. 08007. How can the sun's energy provide a direct source of energy in the future?
4. Make a cross-section model of the earth. Does the earth's interior possess heat? Can we tap this heat energy?
5. Discuss falling water and wind as energy sources.
6. List all of the items used daily that demand electricity or the burning of fuel.
7. List all of the items used daily that demand electricity or the burning of fuel, that you can do without.

With an increasing demand for power, comes a familiar word: pollution.

BIBLIOGRAPHY AND REFERENCES

Barker, Will. *Familiar Reptiles and Amphibians of America.* New York: Harper and Row, Inc., 1964.

Bermant, Roberta M. and Gordon Bermant. "The Behavior of Primates: A Unit for Grades 4 to 6." *American Biology Teacher,* vol. 33, no. 3 (March 1971), pp. 167-171.

Budker, Paul. *The Life of Sharks.* New York: Columbia Press, 1971.

Carolina Biological Supply Company leaflets: "Culture of Amphibian Larvae," "Care and Feeding of Amphibians and Small Reptiles," "Balancing an Aquarium," and "Terrarium Care." Burlington, N.C., Carolina Biological Supply Company.

Carlander, Kenneth D. *Handbook of Freshwater Fishery Biology.* Ames, Iowa: The Iowa State University Press, 1969.

Carr, Archie and the Editors of *Life. The Reptiles.* New York: Time, Inc., 1963.

Carrington, Richard. *The Mammals. (Life Nature Series.)* New York: Time, Inc., 1963.

Cochran, Doris M. *The New Field Book of Reptiles and Amphibians.* New York: Putnam, 1970.

Cooper, Allan. *Fishes of the World.* New York: Bantam Books, Inc., 1971.

Eimerl, Sarel and Irven DeVore. *The Primates. (Life Nature Series.)* New York: Time, Inc., 1965.

Haynes, Thomas M. "An Environmental Study in Biology." *Biological Science Teaching Tips from TST (Science Teacher).* Washington, D.C.: National Science Teachers Association, 1967, pp. 205-206.

Hoffmeister, Donald F. *Mammals of the Grand Canyon.* Chicago: University of Illinois Press, 1971.

Jacobson, W.J. "Population Education." *Ward's Bulletin,* vol. 12, no 86, Rochester, N.Y. (November 1972).

Jernigan, H. Dean. "Care of Living Organisms Makes In-Class Projects." *American Biology Teacher,* vol. 33, no. 7 (October 1971), pp. 412-413.

Linzey, Alicia and Donald. *Mammals of Great Smoky Mountains National Park.* The University of Tennessee Press, 1971.

Lloyd, Glenys and Derk. *Birds of Prey.* New York: Grosset and Dunlap, 1970.

Migdalski, Edward C. *Salt Water Game Fishes.* New York: The Ronald Press Company, 1958.

Orlans, F. Barbara. "The Boundaries of Use of Animals in High School Biology." *Science Teacher,* vol. 35, no. 7 (October 1968), pp. 44-47.

Parker, Bertha M. *Toads and Frogs.* New York: Harper and Row, Inc., 1959.

Peterson, Roger T. and the Editors of *Life. The Birds. (Life Nature Series.)* New York: Time, Inc., 1963.

Smyth, N.R. *Amphibians and Their Ways.* New York: The MacMillan Company, 1963.

The Animal Welfare Institute. *Humane Biology Projects.* New York: The Animal Welfare Institute, 1960.

Turtox Service Leaflets: No. 5, "Starting and Maintaining a Fresh-Water Aquarium"; No. 7, "The Care of Frogs and Other Amphibians"; No. 23, "Feeding Aquarium and Terrarium Animals"; No. 48, "Aquarium Troubles: Their Prevention and Remedies." Chicago: CCM: General Biological, Inc.

Ward's Culture Leaflets: No. 18, "Selection and Use of Aquarium Plants"; No. 19, "Aquaria in the School Laboratory"; and "Use of Terraria in the School Laboratory." Ward's Natural Science Establishment, Inc., Rochester, N.Y.

Walker, Ernest P., *Studying Our Fellow Mammals.* New York: The Animal Welfare Institute, 1967.

Walker, Ernest P. *First Aid and Care of Small Animals.* New York: The Animal Welfare Institute, 1955.

Zim, Herbert and H. Shoemaker. *Fishes.* New York: Golden Press, 1955.

Zim, Herbert and Iran Gabrielson. *Birds.* New York: Golden Press, 1956.

*AUDIO-VISUALS**

AMPHIBIANS

Films:

Amphibians: Frogs, Toads, and Salamanders (BFA)
Frog Development–Fertilization to Hatching (Wards)
From Water to Land (McGraw-Hill)
Life Cycle of a Frog (UW)
Singing Frogs and Toads (IFB)
What Is an Amphibian? (EBEC)
World in a Marsh (McGraw-Hill)

Film Loops:

Eggs and Tadpoles (8 loops) (CCM)
Fish, Amphibians, Reptiles (6 loops) (Holt)

Filmstrips:

Amphibians (EBEC)
Classification of Living Amphibians and Reptiles (4 filmstrips) (EBEC)
Fish and Amphibians (Jam Handy)

FISH

Films:

Environment and Survival: Life in a Trout Stream (BFA)
Fish Are Interesting (BFA)
Fish in a Changing Environment (EBEC)
Reproduction Cycle of Angel Fish (IFB)
The Fish Embryo: From Fertilization to Hatching (EBEC)
What Is a Fish? (EBEC)

Film Loops:

Courtship Ritual of Stickleback Fish (Doubleday)

*Refer to glossary (page 211) for key to abbreviations and addresses.

Film Loops (cont.)

Fish, Amphibians, Reptiles (6 loops) (Holt)
Goldfish Eggs Hatching (Doubleday)
Lungfish and Other Australian Animals (Doubleday)
Reproduction–Fish Eggs Hatchir (Doubleday)
Salmon Run (Doubleday)
Territorial Behavior: Fish (Ealing)

Filmstrips:

Classification of Living Fish (4 filmstrips (EBEC)
Fish and Amphibians (Jam Handy)
Life Cycle of Common Animals (Group I includes the life cycle of the Salmon, parts I and II) (BICO Catalog)

REPTILES

Films:

Reptiles Are Interesting (BFA)
Snakes Are Interesting (IFB)
The Chameleon (IFB)
What Is a Reptile? (EBEC)

Film Loops:

Alligators–Birth and Survival (Doubleday)
Desert Snakes (Doubleday)
Dinosaurs–Meat Eaters (Doubleday)
Rattlesnake (Doubleday)

Filmstrips:

Classification of Living Amphibians and Reptiles (4 filmstrips) (EBEC)
Basic Nature Study: (10 filmstrips including Snakes and Lizards) (BICO Catalog)

BIRDS

Films:

A Wonderful Bird Was the Pelican (United Productions)
Animals That Fly (BFA)
Attracting Birds in Winter (IFB)
Birds of the Sandy Beach: An Introduction to Ecology (BFA)
Birds That Eat Fish (also Insects and Seeds) (IFB)
Building Birdhouses (IFB)
Cultivate Your Garden Birds (IFB)
Flight of the Sea Gulls (IFB)
Prairie Giant (IFB)
Signals for Survival (McGraw-Hill)
Summer Bird Hike (McGraw-Hill)
Waterfowl–A Resource in Danger (EBEC)
Winter Bird Hike (McGraw-Hill)

Film Loops:

American Migratory Birds (Doubleday)
Bird Tricks for Survival (Doubleday)
Birds and Mammals (6 loops) (Holt)

Film Loops (Cont.)

Birds and Their Way of Life (9 loops) (Ealing)
Birds Building Nests (Doubleday)
Courting Rituals of Birds (Doubleday)
Eagle (Doubleday)

Filmstrips:

Audubon's Birds of America (EBEC)
Basic Nature Study (includes birds) (BICO Catalog)

MAMMALS

Films:

Adaptations for Survival: Mammals (IFB)
Animals That Gnaw and Burrow (IFB)
Monkeys (IFB)
Our Endangered Wildlife (McGraw-Hill)
The Cat Family (IFB)
The Dog Family (IFB)
The Horse Family (IFB)
The Living Mammal (IFB)
What Is a Mammal? (EBEC)

Film Loops:

Baboons (Doubleday)
Bats (Doubleday)
Birds and Mammals (6 loops) (Holt)
Bottlenose Dolphin (Doubleday)
Monkeys of the Amazon (Doubleday)
The Honey Glider–Flying Squirrel (Doubleday)
Whales Surfacing (Doubleday)

Filmstrips:

Classification of Living Mammals (6 filmstrips) (EBEC)

VERTEBRATES: LIVING

	BICO	*Carolina*	*Conn.*	*CCM-Turtox*	*So. Bio.*	*Stansi*	*Mogul-Ed*	*Stein.*	*Wards*
AMPHIBIANS									
American Toad	1.00	–	1.25	–	1.25	–	1.35	3.00(6)	–
Marine or giant toad	–	–	–	–	7.50 (4)	–	2.35	7.00(2)	–
Grass frogs (6)	4.50	2.50	4.50	6.55 (12)	4.50 (12)	4.50 (12)	.41 ea.	2.28 (12)	7.50 (12)
Frog eggs, lot	5.50	3.00	2.50	7.05	–	–	4.70	4.00	5.00
Salamanders (3)	4.50	5.00	–	3.55 ea.	–	6.75 (6)	13.55 (12)	6.00 (6)	–

VERTEBRATES: LIVING *(continued)*

	BICO	*Caro-lina*	*Conn.*	*CCM-Tur-tox*	*So. Bio.*	*Stansi*	*Mogul-Ed*	*Stein*	*Wards*
Salamander eggs	–	3.00	–	–	–	–	–	–	5.00
Newts (3)	3.50	4.00	–	1.95 (2)	–	–	4.00 (12)	1.50 (2)	–
Tree frog (6)	–	–	10.00	4.35 ea.	–	–	–	–	–
Bullfrog	–	–	2.00 ea;	2.60 (3)	7.50 (4)	6.75 (3)	2.10 ea.	6.75 (3)	–
Mudpuppy	–	10.00 (3)	–	3.55 ea. ea.	12.50 (6)	6.75 (6)	12.00 (12)	12.00 (12)	24.00 (12)
Amphiuma (Congo eel)	–	8.25	–	3.55	–	–	10.65	–	–
Sirens	–	–	–	–	–	–	4.00 ea.	–	–
REPTILES									
Chameleon (anoli)	.95	1.10	–	2.60 (2)	2.40	–	1.35	8.04 (12)	–
Fence lizard	–	–	–	3.05	2.40	–	–	–	–
Iguana (small)	–	–	–	–	–	–	–	7.00(2)	–
Pond turtles (3"–6")	2.00	1.50	2.25	2.60	8.40 (6)	–	1.00 ea.	1.98 (6)	–
Baby pond turtles (ea.)	1.10	–	1.25	1.55	–	–	2.70	1.50	–
Box turtle (ea.)	–	–	–	3.55	–	–	2.70	1.50	–
Blanding turtle	–	–	–	–	–	–	3.35	1.50	–
Snapping turtle	–	–	–	–	–	–	.50 lb.	.40 lb.	–
Garter snake	–	–	–	–	–	–	2.35	6.00(6)	–
Small non-poisonous snakes (less than 24 in.)		5.00	–	–	–	–	–	–	–
Large non-poisonous snakes (over 24 in.)		7.50	–	–	–	–	–	–	–
FISH									
Mosquito fish *(Gambusia)*	–	3.00(12)	–	–	–	–	–	–	–
Goldfish	–	3.00(6)	–	–	–	–	4.80(12)	4.56(12)	–
Fancy goldfish	–	–	–	–	–	–	8.00(6)	–	–
Medaka egg set	–	5.50	–	–	–	–	–	–	–
Medaka breeding set	–	15.00	–	–	–	–	–	–	–
Zebra fish	–	4.00(6)	–	–	–	–	–	–	–
Mud minnow	–	–	–	–	–	–	3.25(12)	2.52(12)	–
Guppies	–	–	–	–	–	–	3.60(12)	–	–
Brook stickleback	–	–	–	–	–	–	4.80(12)	3.00(12)	–

VERTEBRATES: LIVING *(continued)*

	BICO	*Caro-lina*	*Conn.*	*CCM-Tur-tox*	*So. Bio.*	*Stansi*	*Mogul-Ed*	*Stein.*	*Wards*
MAMMALS									
white rats (6)	–	–	14.00	34.95 (12)	11.30	–	–	1.54 ea.	7.50 (2)
hooded rats (6)	–	–	25.00	4.40 ea.	–	–	–	–	7.50 (2)
white mice (12)	–	–	12.00	20.65	7.50 (6)	–	2.10 ea.	5.00 (2)	7.00 (6)
rabbits	–	–	6.00	–	–	–	–	3.50 ea.	8.50
golden hamsters	–	–	7.00 (12)	8.25 (2)	–	–	2.70	5.00 (2)	8.00 (2)
guinea pigs (cavies)	–	–	12.00(2)	–	–	–	3.00 ea.	2.50 ea.	7.00 ea.
gerbils	–	–	–	10.20 ea.	–	–	6.70 ea.	4.50 ea.	9.00 ea.
BIRDS									
Viable eggs	–	–	6.50 (12)	–	–	–	–	–	6.75 (12)
Pigeons	–	–	15.00 (6)	–	–	–	2.00 ea.	15.00 (12)	–

All prices were in effect at time of writing; check current listings for present offerings and prices. Also note that price differences may also reflect size of specimens, although generally indicate the same approximate size. Also be aware that shipping charges are very high on mammals.

EIGHT

Investigating and Monitoring Pollution

In recent years "pollution" has become a part of our vocabulary. It now touches the air we breathe, the sounds we hear, the water we drink, and the land on which we walk.

While you may be interested in discussing many aspects of "polluting," you may discover that expensive equipment is often necessary to study it seriously.

Nevertheless, there are a variety of simple, inexpensive observations that can complement the more expensive, sophisticated investigations.

SIMPLE WAYS TO OBSERVE AIR POLLUTION

Air pollutants include dust, pollen (Figure 8-1), spores, and other naturally-produced substances as well as man-made materials: particulates to gases, including carbon monoxide, sulfur oxides, nitrogen oxides, and hydrocarbons.

In your studies of air pollution, you may wish to include some of the following:

1. Dust pollen from a flower on a slide containing a drop of water. Examine under the microscope. Using known flowers or tree pollen, describe the various pollen grains; keep a sample for comparative studies. How does pollen affect people?
2. Smear a thin layer of petroleum jelly on a glass slide. Place the slide

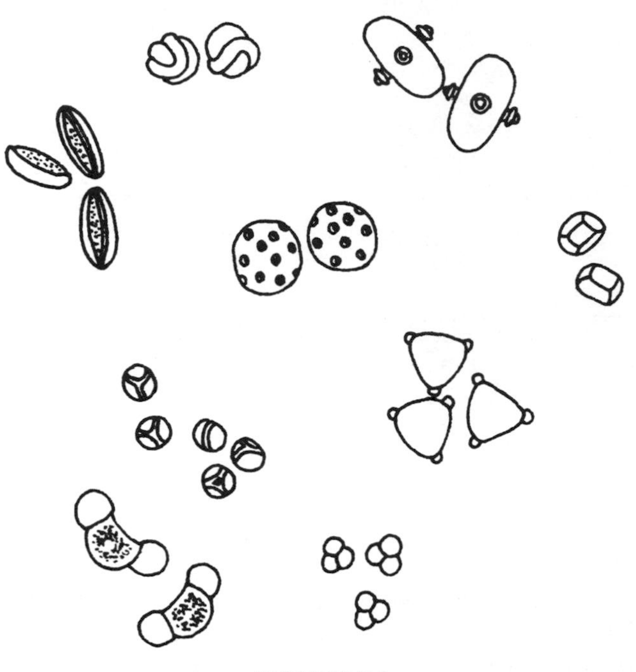

FIGURE 8-1

Pollen exists in a variety of shapes and sizes; man often considers pollen a naturally existing air pollutant.

Original drawings by Sheralyn Lerner; after various authors.

in a partially sheltered, elevated location where it will not be hit directly by wind or rain. Examine the slide under the miscroscope after 24 hours. Conduct pollen counts (rough estimates) and note spores or other particles that settle onto the slide. (Note: a pollen and spore sampling kit is offered by Ward's Natural Science Establishment, Inc., for approximately $80.) At what times in the year is pollen most prevalent? When are spores or dust most prevalent? Graph the results over a period of time. Discuss variables, including wind speed, maturing rates of flowers, nearby building projects, elevation, etc.

3. Cut 6" x 2" strips of poster board and partially fold longitudinally. Punch five holes on one side of each strip and and then fold completely; fasten with tape. Next, carefully stretch Scotch tape

over the bottom of the still exposed holes, with the sticky side facing into the closed portion. Open the strips in a variety of locations for a duration of 24 hours. Later, close them again and seal them until you are ready to examine beneath the microscope. Are more particles in the air at different places in the city?

4. Place cotton pads over the exhausts of a number of automobiles for two minutes while the cars idle. Compare the amount of carbon present, through examination under the dissecting microscope or previous weights compared to weights after the demonstration. Which cars burn the cleanest? What does this indicate in terms of air pollution?
5. Weigh a piece of filter paper and place it outdoors for several days; keep sheltered from rainfall. Once again weigh the paper; what could account for an increase in weight?
6. Make general observations on freshly washed cars. How long does it take for dust to become apparent? What factors control the amount of dust in the atmosphere?
7. Examine a vacuum cleaner filter, placing the dust beneath the microscope. Describe the dust particles. How does the human body remove dust from the atmosphere? What diseases are caused by inhaling large amounts of dust over a period of years?
8. Purchase a Ringlemann scale for approximately $2.00 from Edmund Scientific Company, 615 Edscorp Building, Barrington, New Jersey 08007, or make a scale from whitish gray to black, assigning increasing numbers, with black having the highest value. Compare the scale to smoke emitted from factory smokestacks, etc. Keep records of which stacks emitted the darkest smoke on various days. Compare your data to local or state air pollution data. Do the stacks you have been observing have any effect on the overall air pollution in your area?
9. Purchase a mechanical smoker from Edmund Scientific Company ($10.00) and observe first hand the effects of cigarettes. How much tar and nicotine are created by a pack of burning cigarettes? Test the effects of these products on microscopic organisms, larger invertebrates, or even vertebrates. How do cigarette tars and nicotines affect various organisms?
10. Place a bean or tomato plant in a closed container. Add ozone to the container. How does the presence of ozone affect the plant? Repeat with sulfur dioxide and nitrogen dioxide. If these gases increase with general air pollution, how might plants be affected?
11. Burn a *small* piece of sulfur; describe the odor of the gases emitted. How could these possibly affect humans?

12. Examine two similar wooded habitats, with one in an unpopulated area, and the other near a city or an industrial plant. Compare the animal and plant life in the two locations. Can the absence of organisms (e.g. lichens) indicate a degree of air pollution?
13. Place a bean or tomato plant in a closed container; spray hair spray into the container over a period of weeks. (Do not hit the plant directly with the spray.) Can hair spray components in the atmosphere affect plant life?
14. Purchase a low-cost air pollution test kit from Ward's Natural Science Establishment, Inc. ($16.50) or Edmund Scientific Company ($17.85). Instructions and materials are included for tests of carbon dioxide, carbon monoxide, hydrogen sulfide, nitrogen dioxide, and sulfur dioxide. Test the air around your immediate environment. Compare to tests taken in densely populated and unpopulated areas.
15. You may wish to conduct serious air pollution studies. Ward's Natural Science Establishment, Inc., offers the following materials at these approximate prices:
    a. Educational air pollution study kit, $195
    b. Huey sulfation plates, $44
    c. Gruber particle comparator (sticky paper), $162
    d. Vernamatic balance with air pollution weighing chamber, $570
    e. Universal air tester kit with tubes ($9.50 ea.) for alcohol, ammonia, carbon dioxide, chlorine, hydrogen sulfide, nitrogen dioxide, ozone, and sulfur dioxide, $109

Just as you may choose from a wide range of air pollution investigations, you may also select from a wide range of activities when working with water pollution. However, one specific test has been used extensively in determining whether water is "polluted"; it deals with coliform bacteria.

COLIFORM BACTERIA: A MEASURE OF WATER POLLUTION

Coliform bacteria are organisms that exist in the intestines of humans and other warm-blooded mammals. While they are relatively harmless, they are indicators that more serious, disease-causing bacteria may be nearby. Since the disease-causing bacteria are more difficult to identify, we test for coliform bacteria.

The most significant aspect of coliform bacteria is the number that are present in a given volume of water (100 ml.); a certain amount may be normally present. (The health department in your locality should be able to provide the limits of safety, or numbers that indicate when water is polluted, by their standards.)

To begin, you may want to test simply for the presence of coliform bacteria. This involves adding 1 ml. of a water sample to lactose broth as supplied by Edmund Scientific Company; a color change from purple to green or yellow indicates the presence of the bacteria.

You can make bacterial counts yourself with a variety of methods or kits. You may simply add 1 ml. of a water sample to violet red bile agar and allow 48 hours' incubation at 37°C., or slightly longer at room temperature. You can identify the presence and numbers of bacteria by the iridescent green "sheen" of the coliform bacteria, counting each group or colony as one bacterium. Multiply by 100 to obtain the number of bacteria in 100 ml.

Carolina Biological Supply Company offers an inexpensive kit ($9.75) providing the essentials necessary to conduct total plate counts and differential plate counts of coliform bacteria; this includes student guides.

A more sophisticated method of conducting coliform bacterial counts involves a vacuum filtering apparatus, ampoules of MF-Endo medium, and membrane filters. The necessary equipment may be purchased from the Millipore Corporation, Bedford, Massachusetts. The general directions, after securing the necessary equipment, involve approximately five steps:

1. Sterilize a filter holder with 70 percent isopropyl alcohol or boiling water for three minutes. Load the filter holder with type HA membrane filter (0.45 mm pore diameter), using forceps previously sterilized in alcohol or boiling water.
2. Place an absorbent pad in a 47 mm. Petri dish and break open a 2 ml. ampoule of MF-Endo medium, adding it to the absorbent pad. Cover the dish and set aside temporarily.
3. Place about 20 m. of sterile water in an accompanying funnel that fits over the apparatus in step 1 and mix it with 1 ml. of the sample taken in the environment. Use a vacuum apparatus to force the liquid through the previously loaded membrane filter.
4. Unscrew the funnel and, using sterile forceps, transfer the filter to the absorbent pad prepared in step 2 with the grid side up. Close the Petri dish and invert for 48 hours at room temperature, or 24 hours at 37°C. in an incubator.

5. After the given period of time, use a hand lens to examine the filter paper. Each previously present coliform bacterium will be represented by a colony of bacteria having a greenish sheen. To calculate the number of bacteria in a 100 ml. sample, count the number of colonies and multiply by 100.

Establish counts using various samples. How many colonies in 100 ml. indicate the water is "polluted"? (Ask local health officials for their standards.) Where and when is the water most polluted in your community?

For further information, refer to Leslie and McKinstry (1972) and Sohn and Bullock (1971).

In addition to coliform bacteria, you may wish to test for the presence of oxygen ($6.75 kit from Edmund Scientific Company), detergents, ($5.95 from Edmund Scientific Company), pH, silica, calcium, carbon dioxide, phosphates, nitrates, ammonia, copper, iron, salinity, sulfides, chlorides, chromium, cyanide, dissolved solids, or other substances. (Ward's Natural Science Establishment, Inc. [$99.95], Carolina Biological Supply Company [$34.95], and Sargent Welch, 7300 North Linder Avenue, Skokie, Illinois 60076 [approximately $35], offer kits that cover many of these substances. Edmund Scientific Company has a kit that covers oxygen, carbon dioxide, nitrates, phosphates, total hardness, silica, and pH at six kits for $97.00; they also have an inexpensive water test kit for $12.95.)

A mercury detection kit ($10.95) available from Educational Methods Inc., 500 North Dearborn Street, Chicago, Illinois 60610.

You may also wish to trace stream or river movements with fluorescent dyes; this information could help in understanding the diffusion of pollutants from the sources. (Edmund Scientific Company offers a package of 12 red or yellow/green tablets for $1.50.)

You may wish to clean up streams; in addition to picking up objects and debris, you may wish to purchase large magnets that can remove metal objects from the depths. (Edmund Scientific Company lists a five-pound magnet that can lift up to 150 pounds at a cost of $14.00.)

While chemical tests and the presence of coliform bacteria may indicate water pollution, you may wish to learn more about the type of organisms that frequent polluted water; these can provide further evidence that a body of water is "polluted."

TUBIFEX WORMS AND OTHERS THAT CAN SURVIVE IN POLLUTED WATER

Mayflies and other organisms often indicate a pollution-free environment; the absence of such organisms may be clues that the water is polluted. On the other hand, the presence of pollution-tolerant species may further support your data.

The most familiar pollution-tolerant or pollution-favoring organisms are the *Tubifex* worms. Large mats of *Tubifex* often thrive in the bacteria-laden waters having low oxygen content. It has been suggested that these worms extend their tails from their burrows, with greatest extension when oxygen is lowest in the water.

If an oxygen test is impossible in the field, perhaps you can use the tail extension of *Tubifex* for a general indication of dissolved oxygen. Their presence indicates water pollution and their tail extension, an adaptation to such an environment.

For studies in the classroom, you can order *Tubifex* worms from biological supply houses and maintain them in pond mud on the bottom of an aquarium. If you float a few polyethylene bags of sour cream on the surface, you may help provide an excess of bacteria and low oxygen conditions that *Tubifex* prefer. Refer to Gill (1971) for details.

Other species that may indicate polluted water include:

Algae: (See Figures 8-2 and 8-3.) *Anabaena, Oscillatoria, Euglena gracilis, Chlamydomonas, Scenedesmus quadricauda, Synedra* sp., *Chlorella, Ankistrodesmus falcatus, Phacus pleuronectes, Pandorina marum, Gloeocapsa, Stigeoclonium tenue* (sea lettuce in marine conditions).
Bacteria: *Sphaerotilus*
Fish: Carp
Fungi: *Leptomitus*
Insects: *Chironomus* (bloodworms—see Figure 8-4)
Mollusks: River snails
Protozoa: *Carchesium, Colpidium.*

The presence of algae may complement the study of phosphates; you can purchase any of the listed algae from Carolina Biological Supply Company and experiment with mixing various strengths of detergents or fertilizers in their containers.

You can also use membrane filters for isolating and identi-

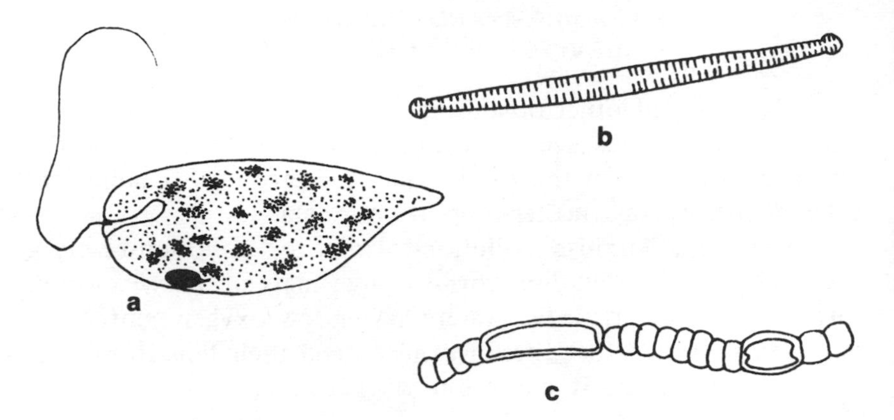

FIGURE 8-2

Three types of algae that may indicate water pollution: (a) *Euglena*, (b) *Synedra* sp., (c) *Anabaena*.

Original drawings by Sheralyn Lerner; after various authors.

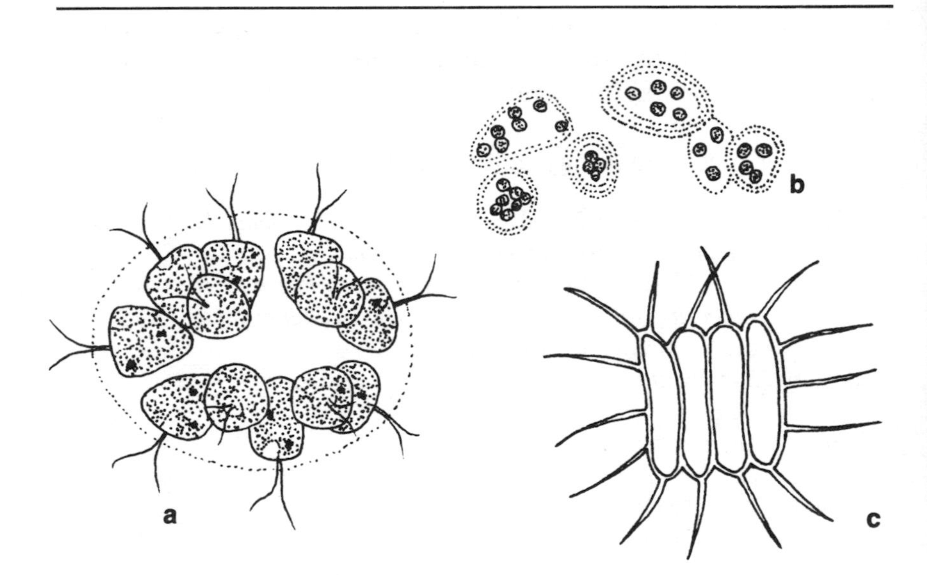

FIGURE 8-3

Three types of algae that may indicate water pollution: (a) *Pandorina*, (b) *Gloeocapsa*, (c) *Scenedesmus*.

Original drawings by Sheralyn Lerner; after various authors.

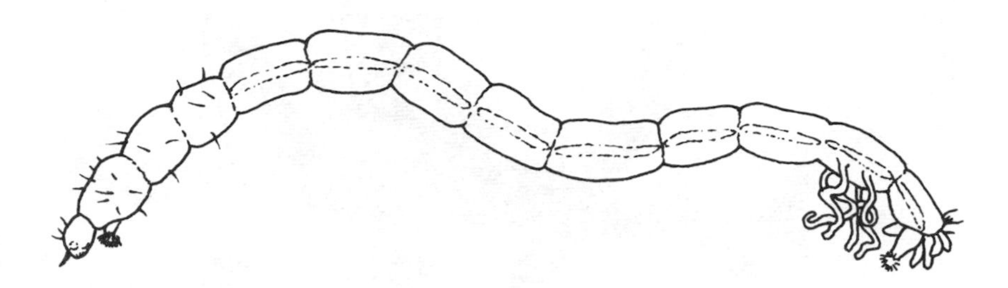

FIGURE 8-4

The bloodworm, *Chironomus,* is an insect that thrives in polluted conditions.

Original drawing by Sheralyn Lerner; after various authors.

---

fying various algae; refer to Sohn (1972) for details on isolating and counting algae on membrane filters. Other references for water pollution include Schlichting (1971) and Abeles (1972) in the bibliography.

Do not forget in your studies that chemical water pollution may be caused by excessive falling of leaves, the death of animals that fall into a water source, etc. There are natural water pollutants to be considered.

Just as air and water pollution have increased with our increasing population and industrialization, so have sounds and noises.

SIMPLE DEMONSTRATION IN NOISE POLLUTION

Noise pollution is becoming an increasing problem in our cities. It is a subtle form of pollution that we may unknowingly accept; loud sounds have become a part of our way of life (Figure 8-5).

Research has shown that noise pollution can cause temporary or permanent loss of hearing and can lead to headaches, muscular tensions, irritability, nausea, decrease in blood flow, less salivary gland and gastric secretion, and more. You may wish to experi-

FIGURE 8-5

Industrialization has added to an increasing amount of noise in our cities.

Photo by Stewart Harris

ment with loud noises; perhaps you can extend man's knowledge on the subject. Your studies might include:

1. Make two 12-inch square containers, one with plywood and the other with soundproofing material. Place an alarm clock in both containers. Which container transmits the loudest sound? How is soundproofing of value in the city? Also discuss and test earplugs.
2. Have students take standard tests. Play loud music while one group takes the test, and permit silence when the other group takes the test. Do loud sounds affect test performance?
3. Play one record repeatedly. When do students request that you stop the sound? How can sounds, even pleasant ones, become a nuisance?

4. Make grating sounds of chalk against the blackboard; record the grinding of teeth, dripping faucets, etc. Do some sounds disturb you even though they may not be loud? How could they influence your daily life?
5. Visit an airport. How are the sounds of airplanes, especially the sonic boom of military aircraft, a problem?
6. Test the distance a sound carries in various locations. Use an alarm clock's alarm and a tape measure to find the distances involved to hear the alarm in a forest, on a street, etc. How do plants (trees and shrubbery) have an important role in noise pollution abatement?
7. Purchase a noise pollution kit (for $24.00) from Educational Methods, Inc., 500 Dearborn Street, Chicago, Illinois 60610; this kit examines the biological and social effects of noise pollution, including simulated hearing losses.
8. Visit a speech and hearing clinic or have a specialist test you for your upper and lower limits of hearing. What is your "threshold of pain"?
9. Purchase a portable noise pollution meter from Edmund Scientific Company, for $76.00. Check your immediate environment for dangerous levels of sound (90-100 dbs and up).

When it comes to garbage and various forms of "land pollution," you may wish to visit local dumps on field trips, but this is one area that may also facilitate the use of role-playing activities.

PROBLEMS OF GARBAGE DISPOSAL: ROLE-PLAYING ACTIVITIES

The following is but one example of a role-playing situation that can be used with pollution topics. It is only as effective as the players and their research; it is intended to reveal complications that enter with "pollution."

Mr. and Mrs. Trash

Mr. and Mrs. Trash live in a small community outside of *[fill in your local city and state]*. They rarely clean their yard and have allowed their trash and garbage to pile up in an adjoining vacant lot.

Only a month ago, the vacant lot was donated to the local Baptist church (which adjoins the property on another boundary) by Mr. Mann, a prominent businessman.

John and Jane Phillips live across the street from the vacant lot and have two children, Sarah (13) and William (12). Yesterday morning Sarah was bitten in the face by a large rat while playing in the vacant lot. Although Dr. Kindle did a professional job in closing the ten-stitch wound, he indicated that a noticeable scar will always remain. In addition, Sarah is just beginning a long series of painful rabies injections.

John and Jane Phillips are naturally upset, but perhaps not as upset as several parties who found they were being sued: namely, Mr. and Mrs. Trash, Mr. Mann, the local Baptist church, and the local health department (rat control division).

Scene 1

Mr. Calhoun, lawyer for the Phillips, calls an informal meeting with the accused parties. He indicates that after an open discussion, his clients (the Phillips) may reconsider their decision to sue; on the other hand, they may concentrate their efforts to sue just a certain individual or lesser number of individuals. Mr. Calhoun leads the discussion:

PARTS

Jane and John Phillips
William and Sarah Phillips
Dr. Kindle
Mary Jones (representing the local Baptist church)
Mr. Mann
Mr. and Mrs. Trash
Mr. Moneyless (chief administrator, representing the health department)
Mr. Calhoun (lawyer for the Phillips)

Scene 2

The Phillips decide to sue. The accused parties, with their lawyers, appear in court before a 12-man jury. The jury must determine if anyone is guilty of negligence; the judge must determine which fines and/or punishments should be administered, or officially dismiss the case. The cases are to be taken individually, in succession, with Mr. Calhoun leading the attack; all official verbiage is heard through lawyers or their witnesses.

PARTS

12-person jury
Mr. Calhoun
Lawyers representing parties being sued (possibly all four present and their clients:

Don Pollit (Mr. Mann's lawyer)
Selma Willson (representing the local Baptist church)
Goomal Pile (representing the health department)
Samuel Refuse (representing Mr. and Mrs. Trash. Also a life-long friend of the family)

Judge (Mrs. Rebecca Stanley—first local woman judge)
Recorder (Miss Binet Stanford)
Jane, John, Sarah, and William Phillips (any others previously connected with the case should also be present)

Scene 3

Regardless of the court outcome, Mr. and Mrs. Phillips feel that the community should be aware of their tragedy and make certain that it does not occur to others. Mr. and Mrs. Phillips appear at a local city council meeting to air their grievances and demand corrections. What do they hear? Look at the city council membership.

PARTS

1. Mayor Dooley (Mayor of the city, and traditionally a politician, presides over the meeting)
2. Miss Sandpiper (a local bird watcher and conservationist)
3. Mr. Sanders (a local crosstown businessman who stands to lose business if the vacant lot is cleaned up; thus far, many of his customers have avoided competitors located near the neighborhood of Mr. and Mrs. Trash)
4. Mr. Lornton (a real estate agent who sells property near Mr. and Mrs. Trash)
5. Dr. Hope (a struggling dentist and lifelong friend of Mr. and Mrs. Trash who resides and works close to them)
6. Mrs. Van Holden (wife of the President of the Junior Chamber of Commerce; often cited for her beautification campaigns in the city)

7. Professor Bridgham (ecology expert who did his doctoral thesis on why man should save rats. His latest newspaper article is titled: "Rats Do Not Bite People; People Bite Rats")
8. Miss Stone (sister of the Union Chief whose crews clean up trashy areas; also, a seeker of causes)
9. Dr. Langehan (once voted on the same issue three different ways; known as the fence swayer. He is near retirement and could care less about anything outside of his Florida home)

You may find it both educational and enjoyable to create your own role-playing activities or simulation games.

SIMULATING POLLUTION IN GAMES

Just as role playing has its advantages in certain situations, simulation games also have their place. They not only make a topic interesting, but they can educate and result in a change in attitudes.

A number of simulation games are offered commercially. "Smog," "Dirty Water," "Population," and "Ecology" are sold by Urban Systems, Inc., 1033 Massachusetts Ave., Cambridge, Mass. 02038. "Extinction" is offered by Sinauer Associates, 20 Second Street, Stanford, Ct. 06905. "The Pollution Game," "The Redwood Controversy," and "Planet Management" can be purchased from Houghton Mifflin Company, 110 Tremont St., Dept. M., Boston, Mass. 02107. Most of these games cost too much for the entire class to play at one time, but one or two games can circulate around the classroom during the year.

Check the shelves of local toy dealers for the latest games that deal with pollution. In addition, you may wish to create your own games as did L.J. Gould in the October, 1972, issue of *Science Teacher* ("The Pollution Game").

It is fitting that our last topic deals with the effects of pollution on wildlife; this combines the emphasis on life and pollution, the two major areas of concern in this book.

EFFECTS OF POLLUTION ON WILDLIFE

While we may show immediate concern for ourselves, we sometimes forget that man is but one species affected by his pollution. Plants can be severely injured by such products as ozone, peroxyacetyl nitrate, nitrogen dioxide, sulfur dioxide, fluoride, or ethylene.

Insecticides meant for pests may accidentally harm birds, bees, and countless other organisms.

Even when man places salt on his streets he introduces a pollutant to countless plants and animals that normally live in salt-free roadside conditions.

Our fertilizers and detergents create excessive algae growth, choking out countless organisms. Our oil spills destroy marine life along the coasts, including numerous birds.

The results of pollution are sometimes best seen by looking for what is not present. Lichens do not normally exist near smoke-filled cities. Certain species of mayflies and dragonflies do not exist near polluted water.

Yet, some species abound in polluted areas, including the bluegreen algae, *Tubifex* worms, carp, and, among others around our cities, rats.

In your studies you may wish to compare and contrast city and country species. You might wish to search for effects of thermal pollution, visual pollution, noise pollution, air pollution, water pollution, and land pollution.

May you continue to discover new aspects about the life around us, and may we all strive to better understand and respect the life on our planet.

BIBLIOGRAPHY AND REFERENCES

Abeles, Ann L. "Phosphate—Some Studies of How It Affects Our Water." *Science Teacher,* vol. 39, no. 2 (February 1972), pp. 53-56.

"Aircraft Noise." *Environmental Science and Technology,* vol. 1 (December 1967), pp. 976-983.

Baron, Robert A. "Noise and Urban Man." *American Journal of Public Health,* vol. 58 (November 1968), pp. 2060-2066.

Blazier, W., "Criteria for Control of Community Noise." *Sound and Vibration,* vol. 2 (May 1968), pp. 516-524.

Breysse, Peter A. "Sound Pollution—Another Urban Problem." *Science Teacher,* vol. 37 (April 1970), pp. 29-34.

Cohen, A., "Industrial and Community Noise Problems and Local Efforts at Control." *Journal of Environmental Health,* vol. 30 (March-April 1968), pp. 516-524.

Cox, George W., editor. *Readings in Conservation Ecology.* New York: Appleton-Century-Crofts, 1969.

Craig, James C. "A Conductivity Device for Measuring Sulfur Dioxide in the Air." *Science Teacher*, vol. 39, no. 9 (December 1972), pp. 42-44.

DeBell, Garrett, editor. *The Environmental Handbook.* New York: Ballantine Books, Inc. 1970.

Evans, Thomas P. "The Secchi Disk: An Instrument for Measuring Water Transparency." *Science Teacher,* vol. 38, no. 4 (April 1971), pp. 57-58.

George, Patricia C. "America's Neglected Pollutant: Solid Waste." *Nation's Cities,* (July 1970).

Gill, Glenn W., "Biostatistical Analysis of *Tubifex* Behavior in Oxygen-Poor Water." *American Biology Teacher,* vol. 33, no. 6 (September 1971), pp. 351-354.

Grissom, John Hoskins. *The Uses and Abuses of Air.* New York: Arno and *The New York Times,* 1970.

Grossman, Mary L. and Shelly, and John M. Hamlet. *Our Vanishing Wilderness.* New York: Madison Square Press, 1969.

Guggisberg, C.A.W. *Man and Wildlife.* New York: Arco Publishing Company, 1970.

Heggestad, H.E. "How Plants Fight 'Man-Made' Pollution." *Science Teacher,* vol. 39, no. 4 (April 1972), pp. 21-29.

Heister, Ralph D., Jr. "The Biotic Index as a Measure of Organic Pollution in Streams." *American Biology Teacher,* vol. 34, no. 2 (February 1972), pp. 79-83.

"Industrial Noise Manual." American Industrial Hygiene Association, Detroit, Michigan, 1966.

Joffe, Joyce. *Conservation.* Garden City, New York: Natural History Press, 1970.

Leslie, Howard C. and Donald M. McKinstry. "Stream Pollution: A Teaching Model." *American Biology Teacher,* vol. 34, no. 2 (February 1972), pp. 99-100.

Man and the Ecosphere: Readings From Scientific American. San Francisco: W.H. Freeman and Company.

McClung, Robert M., *Lost Wild America.* New York: William Morrow and Company, Inc., 1969.

Navarra, John G. *Our Noisy World.* New York: Doubleday and Company, Inc., 1970.

Odum, Howard T. *Environment, Power and Society.* New York: John Wiley and Sons, 1971.

Rondiere, Pierre. *Purity or Pollution: The Struggle for Water.* New York: Franklin Watts, 1971.

Sand, George X. *The Everglades Today: Endangered Wilderness.* New York: Four Winds Press, 1971.

Schlichting, Harold E., Jr. "Measuring Water Quality." *Carolina Tips,* vol. 34, no. 10 (October 1, 1971).

Shurcliff, William A. *SST and Sonic Boom Handbook.* New York: Ballantine Books, Inc., 1969.

Sohn, Bernard I. "Algae as Pollution Indicators: Analysis Using the Membrane Filter." *American Biology Teacher,* vol. 34, no. 1 (January 1972), pp. 19-22.

Sohn, Bernard I. and R. Eugene Bullock. "Bacterial Pollution in Water: Adapting Standard Methods to the Classroom." *American Biology Teacher,* vol. 38, no. 7 (October 1971), pp. 32-35.

Surrarer, T.C. "Pollutants as Input for a Food Chain." *American Biology Teacher,* vol. 34, no. 3 (March 1972), pp. 157-159.

Udall, Stewart L. *The Quiet Crises.* New York: Holt, Rinehart, and Winston, Inc., 1963.

U.S. Department of Agriculture, *Managing Our Environment: A Report on Ways Agriculture Research Fights Pollution.* Washington, D.C., Agriculture Research Service, 1971.

U.S. Department of Agriculture, Science Study Aids No. 5, "Testing for Air Pollution" and No. 4, "Nitrate Water Activities." Beltsville, Maryland: Agriculture Research Center, 1972.

*AUDIO-VISUALS**

Films:

Air of Disaster (U.S. Public Health)
Air Pollution: Take a Deep Deadly Breath (McGraw-Hill)
Aqua Folly (Boyd Films Company)
Ark (Arthur Barr)
Cacophony (United Productions)
Cars in Your Life (McGraw-Hill)
Cities of the Future (McGraw-Hill)
Conservation/For the First Time (McGraw-Hill)
Man's Effect on the Environment (BFA)
New Man in the Forest (IFB)
Our Endangered Wildlife (McGraw-Hill)
People by the Billions (McGraw-Hill)
Poisons, Pests, and People (McGraw-Hill)
Population and Pollution (IFB)
Population Ecology (EBEC)
Population Explosion (McGraw-Hill)
Problems of Conservation–Air (EBEC)
Rise and Fall of the Great Lakes (National Film Board of Canada)
Smoke While You Can (IFB)
Something in the Wind (U.S. Public Health)
Spirit of '76 (Santa Barbara Oil Slick) (American Documentary Films)
The Air Around Us (EBEC)
The Answer Is Clear (General Motors)
The End of One (Learning Corporation of America)
The Garbage Explosion (EBEC)
The House of Man–Part 2, Our Crowded Environment (EBEC)

*Refer to glossary (page 211) for key to abbreviations and addresses; audio-visuals largely apply from upper elementary through high school.

Films (cont.)

The Poisoned Air (U.S. Public Health)
The Problem with Water Is People (McGraw-Hill)
The Refuse Problem (Va. State Dept of Ed.)
The River Must Live (Shell Film Library)
Turn Off Pollution (EBEC)
Wise Use of Water Resources (UW)

Film Loops:

Urban Ecology (10 loops) (Ealing)

Filmstrips:

An Introduction to Human Ecology (6 filmstrips) (Wards)
Ecology and Man (6 filmstrips) (McGraw-Hill)
Environmental Pollution: Our World in Crisis (6 filmstrips) (Wards)
Man-Made World (4 filmstrips) (IFB)
Man's Impact on His Environment (Wards)
Modern Biology: Environment and Survival (BICO Catalog)
The Ecological Crisis (BICO Catalog)
Young Scientists Investigate Pollen (BICO Catalog)

BIOLOGICAL SUPPLY HOUSES

(Write for catalogs)

	Abbreviation
1. BICO Scientific Company Division of National Biological Supply Co. 2325 South Michigan Avenue Chicago, Ill. 60616	BICO
2. Carolina Biological Supply Company Burlington, N.C. 27215 or Powell Laboratories Division Gladstone, Oreg. 92027	CAROLINA
3. CCM: General Biological Inc. (Turtox) 8200 South Hoyne Avenue Chicago, Ill. 60602	CCM-TURTOX
4. Connecticut Valley Biological Supply Co. Valley Road Southampton, Mass. 01073	CONN.
5. J.R. Schettle Biologicals P.O. Box 184 Stillwater, Minn. 55082 or	SCHETTLE (MOGUL-ED)

MOGUL-ED
Oshkosh, Wis. 54901

6. Southern Biological Supply Co. — SO. BIO.
   McKenzie, Tenn. 38201
7. Stansi Scientific Division — STANSI
   Fisher Scientific Company
   1231 North Honore Street
   Chicago, Ill. 60622
8. Steinhilber and Company, Inc. — STEIN. (NASCO)
   or
   NASCO Company
   Fort Atkinson, Wis. 53538
9. Ward's Natural Science Establishment, Inc. — WARDS
   P.O. Box 1712
   Rochester, N.Y. 14603
   or
   P.O. Box 1749
   Monterey, Calif. 93940

SOURCES OF AUDIO-VISUAL AIDS

Listed Name of Supplier	*Full Name and Address*
Arthur Barr	Arthur Barr Productions, Inc. 1029 No. Allen Avenue (P.O. Box 7-C) Pasadena, Calif. 91104
American Documentary	American Documentary Films 379 Bay Street San Francisco, Calif. 94133
Bell	Bell Series American Telephone and Telegraph Co. 195 Broadway New York, N.Y. 10007
BFA	BFA Educational Media 2211 Michigan Avenue Santa Monica, Calif. 90404
BICO Catalog	BICO Scientific Company Chicago, Ill. 60616 (films listed in catalog)

BSCS	Order through: Harcourt, Brace, Jovanovich, Inc. 757 Third Avenue New York, N.Y. 10017 or Houghton Mifflin Company 110 Tremont Street Boston, Mass. 02107 or Rand McNally and Company P.O. Box 7600 Chicago, Ill. 60680
Carolina Biological	Carolina Biological Supply Company Burlington, N.C. 27205
Carousel	Carousel Films Inc. Suite 1503 1501 Broadway New York, N.Y. 10036
CCM	Order through: CCM: Cambosco, Inc. 342 Western Avenue Boston, Mass. 02135 or CCM: General Biological Inc. 8200 South Hoyne Avenue Chicago, Ill. 60620
Coronet	Coronet Films 65 E. South Water Street Chicago, Ill. 60601
Doubleday	Refer to Wards which distributes their film loops.
Ealing	The Ealing Corporation 2225 Massachusetts Avenue Cambridge, Mass. 02140 or Holt Rinehart and Winston, Inc. Media Department 383 Madison Avenue New York, N.Y. 10017

EBEC	Encyclopedia Britannica Educational Corporation Dept. 10A 425 North Michigan Avenue Chicago, Ill. 60611
General Motors	General Motors Corporation 1775 Broadway New York, N.Y. 10019
Holt	Holt, Rinehart and Winston, Inc. Attn: Joseph DiStefano, Media Director 383 Madison Avenue New York, N.Y. 10017
IFB	International Film Bureau 332 S. Michigan Avenue Chicago, Ill. 60604
Indiana University	Indiana University Audio-Visual Center Bloomington, Ind. 47401
Jam Handy	Jam Handy Organization 2821 E. Grand Blvd. Detroit, Mich. 48211
Learning Corporation of America	Learning Corporation of America 711 Fifth Avenue New York, N.Y. 10022
Long Film Service	Long Filmslide Service 7505 Fairmont Avenue Cerrito, Calif. 94530
Martin Moyer	Martin Moyer Productions 900 Federal Avenue Seattle, Wash. 98102
McGraw-Hill	McGraw-Hill Textfilms 330 West 42nd Street New York, N.Y. 10036
Moody Institute	Moody Institute of Science 12000 E. Washington Blvd. Whittier, Calif. 90606

National Film Board of Canada	National Film Board of Canada 680 Fifth Avenue New York, N.Y. 10019
PSP	Popular Science Publishing Co. 239 W. Fairview Blvd. Inglewood, Calif. 90302
Shell Film Library	Shell Film Library 450 North Meridian Street Indianapolis, Ind. 46204
Sterling Films	Sterling Films 43 West 61st Street New York, N.Y. 10023
Sugar Information, Inc.	Sugar Information, Inc. 52 Wall Street New York, N.Y. 10005
SVE	Society for Visual Education, Inc. 1345 Diversey Parkway Chicago, Ill. 60614
Thorne	Thorne Films Dept. GC-71 1229 University Avenue Boulder, Colo. 80302
United Productions	United Productions of America 145 E. 49th Street New York, N.Y. 10017
U.S. Atomic Energy	U.S. Atomic Energy Commission Division of Public Information Washington, D.C. 20545
U.S. Public Health	United States Public Health Service Mr. N.T. Kramis Chief of Graphic Arts Hamilton, Mont. 59840
UW	United World Films (University Education Visual Arts) 221 Park Avenue, S. New York, N.Y. 10003

Va. State Dept. of Ed.	Virginia State Department of Education 523 Eastman Street Richmond, Va. 23219
Wards	Wards Natural Science Establishment, Inc. P.O. Box 1712 Rochester, N.Y. 14603 or P.O. Box 1749 Monterey, Calif. 93940
Walt Disney	Walt Disney Productions Educational Films Divisions 2400 West Almeda Avenue Burbank, Calif. 91500

Index

S

T

U

V